# Mathematical Puzzles

# Mathematical Puzzles:

## A Connoisseur's Collection

Peter Winkler

## CRC Press
Taylor & Francis Group
Boca Raton  London  New York

CRC Press is an imprint of the
Taylor & Francis Group, an **informa** business

CRC Press
Taylor & Francis Group
6000 Broken Sound Parkway NW, Suite 300
Boca Raton, FL 33487-2742

© 2004 by Taylor & Francis Group, LLC
CRC Press is an imprint of Taylor & Francis Group, an Informa business

No claim to original U.S. Government works

ISBN-13: 9781568812014 (pbk)

**Visit the Taylor & Francis Web site at**
**http://www.taylorandfrancis.com**

**and the CRC Press Web site at**
**http://www.crcpress.com**

To Lois, for love and support; to Martin Gardner, for inspiration; and to the many misguided friends, relatives, colleagues, and reviewers who are praying that the publication of this book will bring about a diminution in my passion for puzzles.

# Contents

# Preface

Doubt is the vestibule which all must pass before they can enter the temple of wisdom. When we are in doubt and puzzle out the truth by our own exertions, we have gained something that will stay by us and will serve us again. But if to avoid the trouble of the search we avail ourselves of the superior information of a friend, such knowledge will not remain with us; we have not bought, but borrowed it.

—*C. C. Colton*

These puzzles are not for everybody.

To appreciate them, and to solve them, it is necessary—but not sufficient—to be comfortable with mathematics. You will need to know what a point and a line are, what a prime number is, and how many ways there are to arrange the five cards in a poker hand. Most importantly, you will need to know what it means to *prove* something.

You will *not* need a professional acquaintance with mathematics. You know what a group is? Fine—but you won't need it here. Your computer, calculator, and calculus text can stay in their boxes; but your thinking-cap will have to be on.

Who are you? Amateur mathematicians. Scientists of all kinds. Bright high school and college students. And yes, professional mathematicians and teachers of mathematics will discover new challenges here. These puzzles are not (usually) found in journal articles, in homework exercise lists, or in other puzzle books.

So where did I get them? Word of mouth. Among mathematicians, puzzles like these spread the way jokes spread. In some cases, I have been able to trace the puzzle to a written source, such as an All Soviet Union Mathematical Competition, an International Mathematics Olympiad, or a Martin Gardner column, but of course, that isn't necessarily the original written source, and even if it were, some form of the puzzle might have been bandied about orally for years first. In a few cases, I can name the puzzle's inventor (e.g., when I devised it myself). Often the solution is my own and not necessarily the one intended by the composer; multiple solutions are presented only when they are irresistible.

The wording of the puzzles, and their solutions, is my own and I must take full responsibility for errors and ambiguities. Feel free to send complaints, corrections, and puzzle source information via email to pw@akpeters.com. (An exception: As noted in Chapter 12, I am not the right person to whom to send proposed solutions to Unsolved Puzzles.)

As of this writing, I have been a professional mathematician for 28 years (14 in academia, 14 in industry) and have collected mathematical puzzles since my own high school days in the '60s. What you see here are only my favorite hundred or so puzzles. To make it into this book, a puzzle should satisfy most of the following criteria.

Amusement: It should be entertaining. Problems on the William Lowell Putnam Mathematical Competition, given annually to college students in the US and Canada, are devised to test the students' ability; that is a fine objective, but not always consistent with amusement. (There *are* a few puzzles from the Putnam in this book, however.)

Universality: It should suggest some general mathematical truth. Complex logic puzzles, algebraic puzzles of the type "In two years, Alice will be twice as old as Bob was when...," puzzles relying on properties of particular large numbers, and many other kinds of cleverly devised problems are ruled out.

Elegance: It should be elementary and easy to state. After all, to be passed orally it must be easy to remember! If the statement carries an element of surprise, so much the better.

Difficulty: It should not be obvious how to solve the puzzle.

Solvability: It should boast at least one solution which is elementary and easily convincing.

The last two points create a tension: The puzzle should have an easy solution, yet not be easy to solve. Like a good riddle, the answer should be hard to find, but easy to appreciate. Of course, in the case of the Unsolved Puzzles in Chapter 12, the difficulty is evident and the last constraint must be forgiven.

A word on format. The puzzles are organized into chapters for convenience, classified loosely by mathematical area of statement or solution. The solutions are presented at the end of each chapter (except the last); the end of each solution is marked with a heart symbol ($\heartsuit$). If there is information about the background and source of a puzzle, it is presented here. A puzzle's statement is *not* repeated at the head of its solution; I want to encourage readers to tackle all the puzzles in each chapter before reading the answers.

These puzzles are hard. Several existed as unsolved problems before someone came up with the (elegant) solution you will read here. The Unsolved Puzzles at the end of the book are thus a logical wind-up to the collection, perhaps only slightly harder than the others.

You can take pride in any puzzles you solve, and even more in any for which you find better solutions than mine.

Good luck!

*—Peter Winkler*

# Insight

> During [these] periods of relaxation after concentrated intellectual activity, the intuitive mind seems to take over and can produce the sudden clarifying insights which give so much joy and delight.
>
> —*Fritjof Capra, physicist*

This warm-up chapter contains a variety of puzzles not associated with a particular topic or technique. As is often the case, however, some key insight will put you on the right track. Here's one to get you started:

## Coins in a Row

On a table is a row of fifty coins, of various denominations. Alice picks a coin from one of the ends and puts it in her pocket; then Bob chooses a coin from one of the (remaining) ends, and the alternation continues until Bob pockets the last coin.

Prove that Alice can play so as to guarantee at least as much money as Bob.

Try this yourself with some coins (or random numbers), perhaps just 4 or 6 of them instead of 50; it's not obvious how best to play, is it? But then, maybe Alice doesn't need the *best* strategy. Here's your chance to set a precedent for yourself by trying to solve this one before reading further.

Solution: Number the coins from 1 to 50 and observe that no matter how Bob plays, Alice can capture all the even-numbered coins, or, if she prefers, all the odd-numbered coins. One of these choices must at least match the other.                    ♡

This puzzle, passed to me by mathematician Ehud Friedgut, was alleged to have been used by a high-tech company in Israel to test job candidates. In general Alice has even better strategies than choosing all the even or all the odd coins. However, if there are 51 coins instead of 50, it is usually Bob (the second to play) who will have the advantage, despite collecting fewer coins than Alice. It seems paradoxical that the parity of the number of coins has such a huge effect on the outcome of this game, in which all of the action takes place at the ends.

You're on your own now. We'll begin with two puzzles which are a bit less mathematical, then move on to the more serious stuff. Let your imagination be your guide!

# The Bixby Boys

It was the first day of class and Mrs. Feldman had two identical-looking pupils, Donald and Ronald Bixby, sitting together in the first row.

"You two are twins, I take it?" she asked.

"No," they replied in unison.

But a check of their records showed that they had the same parents and were born on the same day. How could this be?

# The Attic Lamp Switch

A downstairs panel contains three on-off switches, one of which controls the lamp in the attic—but which one? Your mission is to do something with the switches, then determine after *one* trip to the attic which switch is connected to the attic lamp.

# Gasoline Crisis

There's a gasoline crisis, and the fuel stations located on a long circular route together contain just enough gas to make one trip around. Prove that if you start at the right station with an empty tank, you can make it all the way around.

# Uses of Fuses

You are presented with two fuses (lengths of string), each of which will burn for exactly 1 minute, but not uniformly along its length. Can you use them to measure 45 seconds?

# Integers and Rectangles

A large rectangle in the plane is partitioned into smaller rectangles, each of which has either integer height or integer width (or both). Prove that the large rectangle also has this property.

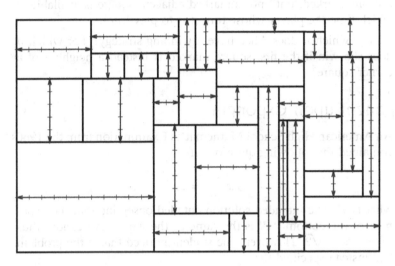

# Tipping the Scale

A balance scale sits on the teacher's table, currently tipped to the right. There is a set of weights on the scales, and on each weight is the name of at least one pupil. On entering the classroom, each pupil moves all the weights carrying his or her name to the opposite side of the scale. Prove that there is *some* set of pupils that you, the teacher, can let in which will tip the scales to the left.

# Watches on the Table

Fifty accurate watches lie on a table. Prove that there exists a moment in time when the sum of the distances from the center of the table to the ends of the minute hands is more than the sum of the distances from the center of the table to the centers of the watches.

# Path on a Chessboard

Alice begins by marking a corner square of an $n \times n$ chessboard; Bob marks an orthogonally adjacent square. Thereafter, Alice and Bob continue alternating, each marking a square adjacent to the last one marked, until no unmarked adjacent square is available at which time the player whose turn it is to play loses.

For which $n$ does Alice have a winning strategy? For which $n$ does she win if the first square marked is instead a neighbor of a corner square?

# Exponent upon Exponent

An American High School Mathematics Examination from the 1960s contained the following question: If

$$x^{x^{x^{\cdot^{\cdot^{\cdot}}}}} = 2,$$

what is $x$? The intended solution entailed observing that the exponent of the bottom "$x$" is the same as the whole expression, thus $x^2 = 2$, $x = \sqrt{2}$. However, one student noticed that if the problem had instead specified that

$$x^{x^{x^{\cdot^{\cdot^{\cdot}}}}} = 4,$$

he would have obtained the same answer: $x = \sqrt[4]{4} = \sqrt{2}$.

Hmm. Just what is $\sqrt{2}^{\sqrt{2}^{\sqrt{2}^{\cdot^{\cdot^{\cdot}}}}}$, anyway? Can you prove it?

# Soldiers in the Field

An odd number of soldiers are stationed in a field, in such a way that all the pairwise distances are distinct. Each soldier is told to keep an eye on the nearest other soldier.

Prove that at least one soldier is not being watched.

# Intervals and Distances

Let $S$ be the union of $k$ disjoint, closed intervals in the unit interval $[0, 1]$. Suppose $S$ has the property that for every real number $d$ in $[0, 1]$, there are two points in $S$ at distance $d$. Prove that the sum of the lengths of the intervals in $S$ is at least $1/k$.

# Summing to 15

Alice and Bob alternately choose numbers from among 1, 2, ..., 9, without replacement. The first to obtain 3 numbers which sum to 15 wins. Does Alice (the first to play) have a winning strategy?

# Solutions and Comments

## The Bixby Boys

A classic brain-teaser. They were triplets, of course. The third (Arnold?) was in another class. ♡

## The Attic Lamp Switch

This puzzle swept the world like a flu epidemic about a decade ago; I don't know its original source.

It really is impossible to tell which switch is connected to the attic lamp if all you get is one bit of information from your trip to the attic. However, you can get more information with your hand! Turn on Switches 1 and 2, wait a few minutes, then turn off Switch 2 before ascending to the attic. If the bulb is off, but warm, you conclude that Switch 2 is the winner. ♡

If you can't reach the bulb, but have *enormous* patience, you can achieve the same effect by turning Switch 2 on and then waiting a couple of months before turning on Switch 1 and visiting the attic. If the bulb is burnt out, Switch 2 is the culprit.

## Gasoline Crisis

This puzzle has been around for a long time and can be found, for example, in László Lovász's marvelous book, *Combinatorial Problems and Exercises*, North Holland, Amsterdam, 1979. The trick is to imagine that you begin at Station 1 (say) with *plenty* of fuel, then proceed around the route, emptying each station as you go. When you return to Station 1, you will have the same amount of fuel in your tank as when you started.

As you do this, keep track of how much fuel you have left as you pull into each station; suppose that this quantity is minimized at Station $k$. Then, if you start at Station $k$ with an empty tank, you will not run out of fuel between stations. ♡

## Uses of Fuses

Simultaneously light both ends of one fuse and one end of the other; when the first fuse burns out (after half a minute), light the other end of the second. When it finishes, 45 seconds have passed.♡

This and other fuse puzzles seem to have spread like wildfire a few years ago. Recreational mathematics expert Dick Hess has put together a miniature volume called *Shoelace Clock Puzzles* devoted to them; he first heard the one above from Carl Morris of Harvard University.

Hess considers multiple fuses (shoelaces, for him) of various lengths, but lights them only at ends. If you allow midfuse ignition and arbitrary dexterity, you can do even more. For example, you can get 10 seconds from a single 60-second fuse by lighting at both ends and at two internal points, then lighting a new internal point every time a segment finishes; thus, at all times, three segments are burning at both ends and the fuse material is being consumed at six times the intended rate.

Bit of a mad scramble at the end, though. You'll need infinitely many matches to get perfect precision.

## Integers and Rectangles

This puzzle was the subject of a unique article by Stan Wagon (of Macalester College in St. Paul, MN) called "Fourteen Proofs of a Result about Tiling a Rectangle," in *The American Mathematical Monthly*, Vol. 94 (1987), pp. 601–617.

Some of Wagon's solutions make amusing use of heavy mathematical machinery; one that doesn't entails placing the lower left-hand corner of the big rectangle on the origin of a grid made up of squares of side $1/2$. Coloring the grid squares alternately black and white, as on a chessboard, we see that each small rectangle is exactly half white and half black. The same, consequently, is true for the big rectangle. However, if (say) the height of the big rectangle is not integral, the region of the big rectangle between the lines $x = 0$ and $x = 1/2$ will not be color-balanced. Hence, the width would have to be integral. ♡

Your author is responsible for the following solution, not found in Wagon's article. Letting $\varepsilon$ be less than the smallest tolerance in the partition, color each small rectangle of integral width green except for a red horizontal strip of width $\varepsilon$ across the top, and another across the bottom. Color each remaining small rectangle red, except for a green vertical strip of width $\varepsilon$ along the left side and another along the right.

Place the lower left point of the big rectangle at the origin. Either there is a green path from the left side of the big rectangle to the right side, or a red path from bottom to top; suppose the former. Every time the green path crosses a vertical border of the partition, it is at an integral coordinate; thus, the big rectangle has integral width. Similarly a red path from bottom to top forces integral height.

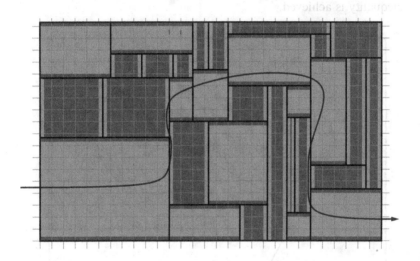

Tipping the Scale

Consider all subsets of students, including the empty set and the full set. Each weight will be on the left half the time, so the total weight on the left for all these subsets is the same as the total weight on the right. Since the empty set results in a tip to the right, some other set must tip to the left.

(*Source*: Second All Soviet Union Mathematical Competition, Leningrad 1968). ♡

The "averaging" technique used here comes up often: watch for it!

### Watches on the Table

Considering just one watch, we claim that during the passing of an hour, the average distance from the center $C$ of the table to the tip $M$ of the minute hand exceeds the distance from $C$ to the center $W$ of the watch. This is so because if we draw a line $L$ through $C$ perpendicular to the line from $C$ to $W$, then the average distance from $L$ to $M$ is clearly equal to $LW$ which is in turn equal to $CW$. But $CM$ is at least equal to $LM$ and usually bigger.

Of course, if we sum over all of the watches we reach the same conclusion, and it follows that sometime during the hour the desired inequality is achieved. ♡

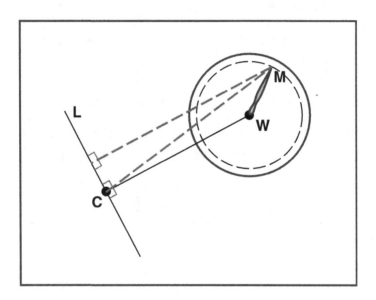

The requirement that the watches be accurate is to ensure that each minute hand moves at constant speed. It doesn't matter if those speeds differ, unless our patience is limited to one hour.

One additional note: If you set and place the watches carefully, you *can* ensure that the sum of the distances from the center of the table to the ends of the minute hands is always strictly greater than the sum of the distances from the center of the table to the center of the watches.

*Source:* This puzzle appeared in the 10th All Soviet Union Mathematical Competition, Dushanbe, 1976.

## Path on a Chessboard

If $n$ is even, Bob has a simple winning strategy no matter where Alice starts. He merely imagines a covering of the board by dominoes, each domino covering two adjacent squares of the board. Bob then plays in the other half of each domino started by Alice. (Note that this works for Bob even when Alice is allowed to mark any square she wants at each move!)

When $n$ is odd, and Alice begins in a corner, she wins by imagining a domino tiling that covers every square except the corner in which she starts.

However, Alice loses in the odd $n$ case when she must start in the square adjacent to the corner. Suppose the corner squares are black in a checkerboard coloring, so that her starting square is white. There is a domino tiling of the whole board minus one black square; Bob wins by completing these dominoes. Alice can never mark the one uncovered square because all the squares she marks are white.♡

*Source:* The 12th All Soviet Union Mathematical Competition, Tashkent, 1978.

## Exponent upon Exponent

If it means anything at all,

$$\sqrt{2}^{\sqrt{2}^{\sqrt{2}^{\cdot^{\cdot^{\cdot}}}}}$$

is the limit of the sequence $\sqrt{2}$, $\sqrt{2}^{\sqrt{2}}$, $\sqrt{2}^{\sqrt{2}^{\sqrt{2}}}$, .... . In fact, the limit does exist; the sequence is increasing and bounded above.

To show the former, we name the sequence $s_1, s_2, \ldots$ and prove by induction that $1 < s_i < s_{i+1}$ for each $i \geq 1$. This is easy because $s_{i+2} = \sqrt{2}^{s_{i+1}} > \sqrt{2}^{s_i} = s_{i+1}$.

To get the bound, observe that if we replace the top $\sqrt{2}$ in any $s_i$ by the larger value 2, the whole expression collapses to 2.

Now that we know the limit exists, let us call it $y$; it must indeed satisfy $\sqrt{2}^y = y$. Looking at the equation $x = y^{1/y}$, we observe using elementary calculus (oops—sorry) that $x$ is strictly increasing in $y$ up to its maximum at $y = e$ and strictly decreasing thereafter. Thus, there are at most two values of $y$ corresponding to any given value of $x$, and for $x = \sqrt{2}$, we know the values: $y = 2$ and $y = 4$.

Since our sequence is bounded by 2, we rule out 4 and conclude that $y = 2$. ♡

Generalizing the above argument, we see that $x^{x^{x^{\cdot^{\cdot^{\cdot}}}}}$ is meaningful and equal to the lower root of $x = y^{1/y}$, as long as $x \leq e^{1/e}$. For $x = e^{1/e}$, the expression is equal to $e$, but as soon as $x$ exceeds $e^{1/e}$, the sequence diverges to infinity.

## Soldiers in the Field

This problem, from the 6th All Soviet Union Mathematical Competition in Voronezh, 1966, is most easily solved by considering the two soldiers at shortest distance from each other. Each of these watches the other; if anyone else is watching one of them, then we have a soldier being watched twice and therefore another not being watched at all. Otherwise, these two can be removed without affecting the others. Since the number of soldiers is odd, this procedure would eventually reduce to one soldier not watching anyone, a contradiction. ♡

## Intervals and Distances

*Source:* This puzzle comes from the 17th All Soviet Union Mathematical Competition, Kishenew, 1983.

Suppose the lengths of the intervals in $S$ are $s_1, \ldots, s_k$, summing to $s$. Let us consider the interval $I_{ij}$ of *distances* obtainable by taking one point from the $i$th interval and one from the $j$th. Clearly, $I_{ij}$ has length $s_i + s_j$. Summing over all pairs of intervals, each $s_i$ appears $k-1$ times, so the total measure of the distances obtainable by choosing points from two different intervals is at most $(k-1)s$. Distances obtainable by taking two points from the *same* interval run from 0 only up to the maximum of the lengths $s_i$, so altogether the measure of the distances is at most $ks$; from $ks \geq 1$, we get $s \geq 1/k$. ♡

The argument is tight only if the maximum $s_i$ is equal to $s$, i.e., all the intervals but one have length 0. This we can achieve, by taking one interval to be $[j/k, (j+1)/k]$ for some $j \in \{0, 1, \ldots, k-1\}$, and adding the single points 0, $1/k$, $2/k$, $\ldots$, $(j-2)/k$, $(j-1)/k$, $(j+2)/k$, $(j+3)/k$, $\ldots$, 1.

## Summing to 15

The quick way to solve the puzzle is to imagine that Alice and Bob are playing on the following magic square:

$$\begin{matrix} 8 & 1 & 6 \\ 3 & 5 & 7 \\ 4 & 9 & 2 \end{matrix}$$

Since it is exactly the rows, columns, and main diagonals which sum to 15, they are playing Tic-Tac-Toe! Everyone knows that best play in Tic-Tac-Toe leads to a draw, so the answer to our question is no, Alice does not have a winning strategy.

This silly game is mentioned in Vol. II of the classic *Winning Ways for Your Mathematical Plays* by Elwyn Berlekamp, John Conway and Richard Guy (Academic Press, 1982; 2nd Edition, A K Peters 2001). The book attributes the puzzle to one E. Pericoloso Sporgersi, but, rather suspiciously, the phrase is found also on Italian railroad trains, warning passengers not to lean out the window.

# Numbers

We learned to be happy
We danced 'round the hall
And learning to count was the key to it all.

*---The Count, "Sesame Street"*

Numbers are an endless source of fascination, and for some, a life-long disease. Some people can be captured even by the properties of *particular* numbers; many intriguing puzzles have been concocted to take advantage of special properties, often by requiring deductions from what appears to be a shortage of data.

The spirit of this collection, however, suggests striving for greater universality. Our number-theoretic problems are about numbers in general, not particular ones. In most cases, you will need little more to solve them than the fact that every positive integer is uniquely expressible as the product of powers of primes.

Here is a practice problem:

## Locker Doors

Lockers numbered 1 to 100 stand in a row at the school gym. When the first student arrives, she opens all the lockers. The second student then goes through and recloses all the even-numbered lockers; the third student changes the state of every locker whose number is a multiple of 3.

This continues until 100 students have passed through. Which lockers are now open?

Solution: The state of locker $n$ is changed when the $k$th student passes through, for every divisor $k$ of $n$. Here, we make use of the fact that divisors *usually* come in pairs $\{j, k\}$ where $j \cdot k = n$; so the net effect of students $j$ and $k$ on this locker is nil. The exception is when $n$ is a perfect square, in which case there is no other divisor to cancel the effect of the $\sqrt{n}$th student; therefore, the lockers which are open at the end are exactly the perfect squares, 1, 4, 9, 16, 25, 36, 49, 64, 81, and 100. ♡

We'll start with a couple of observations about base 10 representation of integers, and finish with a surprisingly subtle dinner table conundrum.

# Zeroes, Ones, and Twos

Let $n$ be a natural number. Prove that (a) $n$ has a (nonzero) multiple whose representation (base 10) contains only zeroes and ones; and (b) $2^n$ has a multiple whose representation contains only ones and twos.

# Sums and Differences

Given 25 different positive numbers, prove that you can choose two of them such that none of the other numbers equals either their sum or their difference.

# Generating the Rationals

A set $S$ contains 0 and 1, and the mean of every finite nonempty subset of $S$. Prove that $S$ contains all the rational numbers in the unit interval.

# Summing Fractions

Given a natural number $n > 1$, add up all the fractions $1/pq$, where $p$ and $q$ are relatively prime, $0 < p < q \le n$, and $p + q > n$. Prove that the result is always $1/2$.

# Subtracting around the Corner

Write a sequence of $n$ positive numbers. Replace each by the absolute difference between it and its successor (going around the corner). Repeat until all the numbers are 0. Prove that for $n = 5$ the process may go on forever, but for $n = 4$ it always terminates.

# Profit and Loss

At the stockholders' meeting, the board presents the month-by-month profits (or losses) since the last meeting. "Note," says the CEO, "that we made a profit over every consecutive eight-month period."

"Maybe so," a shareholder complains, "but I also see we *lost* money over every consecutive *five*-month period!"

What's the maximum number of months that could have passed since the last meeting?

# First Odd Number in the Dictionary

Each number from 1 to $10^{10}$ is written out in formal English (e.g., "two hundred eleven," "one thousand forty-two") and then listed in alphabetical order (as in a dictionary, where spaces and hyphens are ignored). What's the first odd number in the list?

# Solutions and Comments

### Zeroes, Ones, and Twos

For Part (a), we employ the famous and useful "pigeon-hole principle": If there are more pigeons than holes, then some hole must contain at least two pigeons. There are only $n$ numbers modulo $n$, but the set $\{1, 11, 111, 1111, \dots\}$, whose largest member has $n+1$ digits, has size $n+1$; thus, it contains two numbers whose values are equal modulo $n$. Subtract one from the other!　　♡

As pointed out to me by David Gale, as long as $n$ is not a multiple of 2 or 5, you can even find a multiple of $n$ whose representation (base 10) is all ones. The reason is that the above argument produces a multiple of $n$ of the form $111\dots111000\dots000$; if there are $k$ zeros at the end, dividing by $10^k = 2^k \cdot 5^k$ leaves you with all ones and still with a multiple of $k$.

For Part (b), it's perhaps easiest to show by induction on $k$ that there is a $k$-digit number containing only ones and twos which is a multiple of $2^k$. Adding a 1 or a 2 to the front of such a number increments it by $2^k 5^k$ or by $2^{k+1}5^k$, in each case preserving divisibility by $2^k$; since the two choices differ by $2^k 5^k$, one of them must actually achieve divisibility by $2^{k+1}$.　　♡

The first of these problems came to me from Muthu Muthukrishnan, of AT&T Research and Rutgers University. The second appeared on the 5th All Soviet Union Mathematical Competition, Riga, 1971; the solution given here was suggested by Sasha Barg of the University of Maryland.

A similar problem, from the 1st All Soviet Union Mathematical Competition, Tbilisi, 1967, asks for a proof that there exists a number divisible by $5^{1000}$ not containing any 0 in its decimal representation. One approach is to assume otherwise and let $k$ be the biggest power of 5 available. Let $n$ have the most factors of 5 (say, $j \le k$) of any number with at least $k$ digits, none of which is a zero. Then if $n \equiv d \mod 5^{j+1}$, subtracting $d \cdot 10^{j+1}$ or adding $(5-d) \cdot 10^{j+1}$ will keep $n$ zero-free and boost its divisibility, a contradiction. $\heartsuit$

### Sums and Differences

This problem also appeared on the 5th All Soviet Union Mathematical Competition, Riga, 1971. Let the numbers be $x_1 < x_2 < \cdots < x_n$. If $x_n$ is unavailable to be taken as one of the desired numbers, it must be that for each lower number $x_i$, there is another $x_j$ with $x_i + x_j = x_n$. Thus, the first 24 numbers are paired in such a way that $x_i + x_{n-i-1} = x_n$. Now consider $x_{n-1}$ together with any of $x_2, \ldots, x_{n-2}$; these pairs sum to more than $x_n = x_{n-1} + x_1$ and so $x_2, \ldots, x_{n-2}$ must also be paired, this time satisfying $x_{2+i} + x_{n-2-i} = x_{n-1}$. But that leaves $x_{(n-1)/2}$ paired with itself, so the numbers $x_{n-1}$, $x_{(n-1)/2}$ solve the problem. $\heartsuit$

### Generating the Rationals

First note that $S$ contains all the "dyadic" rationals, that is, rationals of form $p/2^n$; we can obtain all those with denominator $2^n$ and odd numerator by averaging two adjacent ones with lower-powered denominators.

Now any general $p/q$ is of course the average of $p$ ones and $q-p$ zeros. We choose $n$ large and replace the zeros by $1/2^n$, $-1/2^n$, $2/2^n$, $-2/2^n$, $3/2^n$, etc. including one 0 if $p$ is odd. Similarly, we replace the ones by $1 - 1/2^n$, $1 + 1/2^n$, $1 - 2/2^n$ and so forth. Of course, some of these numbers lie outside the unit interval, but we can rescale the procedure to fit some dyadic interval containing $p/q$ and lying strictly between 0 and 1. $\heartsuit$

*Source:* The 13th All Soviet Union Mathematical Competition, Tbilisi, 1979.

### Summing Fractions

We proceed by induction, noting that the statement is true for $n = 2$. Moving from $n$ to $n+1$, you gain $1/pn$ for each $p$ with $(p, n) = 1$, and lose $1/pq$ for each $p$ and $q$ with $(p, q) = 1$ and $p + q = n$. Thus, each pair $p, q$ satisfying the conditions of the puzzle signifies a loss of $1/pq$ but a gain of $1/pn + 1/qn = 1/pq$, neatly cancelling. ♡

*Source:* The 3rd All Soviet Union Mathematical Competition, Kiev, 1969.

### Subtracting around the Corner

A substitute high school math teacher (at Fair Lawn Senior High School, New Jersey, 1962) told me that some WWII prisoner of war entertained himself by trying various sequences of four numbers to see how long he could get them to survive under the above operations.

Considering values modulo 2 solves both problems. In the $n = 4$ case, up to rotation and reflection, 1 0 0 0 and 1 1 1 0 become 1 1 0 0, then 1 0 1 0, then 1 1 1 1, then 0 0 0 0. Since this covers all cases, we see that when working with ordinary integers, at most four steps are required to make all the numbers even; at that point, we may as well divide out by the largest common power of two before proceeding. Since the value $M$ of the largest number in the sequence can never increase, and drops by a factor of 2 or more at least once every four steps, the sequence must hit 0 0 0 0 after at most $4(1 + \lceil \log_2 M \rceil)$ steps.

On the other hand, for $n = 5$, the sequence 1 1 0 0 0 (considered either as binary or ordinary numbers) cycles via 1 0 1 0 0, 1 1 1 1 0, 1 1 0 0 0. ♡

A little analysis via polynomials over the integers modulo 2 shows that the salient issue is whether $n$ is a power of 2.

### Profit and Loss

This puzzle is adapted from one which appeared on the 1977 International Mathematical Olympiad, submitted by a Vietnamese composer; thanks to Titu Andreescu for telling me about it. The solution below is my own, however.

What's needed, of course, is a maximum-length sequence of numbers such that every substring of length 8 adds up to more than

0, but every substring of length 5 adds up to less than 0. The string must certainly be finite, in fact less than 40 in length, else you could express the sum of the first 40 entries both as the (positive) sum of 5 substrings of length 8 and the (negative) sum of 8 substrings of length 5.

Let's tackle the problem more generally and let $f(x, y)$ be the length of the longest string such that every $x$-substring has positive sum and every $y$-substring negative; we may suppose $x > y$. If $x$ is a *multiple* of $y$, then $f(x, y) = x - 1$ and we must accept vacuous truth with respect to the $x$-substrings.

What if $y = 2$ and $x$ is odd? Then you can have a string of length $x$ itself, with entries that alternate between, say, $x - 1$ and $-x$. But you can't have $x + 1$ numbers, because in each $x$-substring the odd entries must be positive (since you can cover it with 2-substrings leaving out any odd entry). But there are two $x$-substrings and together they imply that the middle two numbers are both positive, a contradiction.

Applying this reasoning more generally suggests that $f(x, y) \leq x + y - 2$ when $x$ and $y$ are relatively prime, i.e., they have no common divisor other than 1. We can prove this by induction as follows. Suppose to the contrary that we have a string of length $x + y - 1$ which satisfies the given conditions. Write $x = ay + b$ where $0 < b < y$, and look at the last $y + b - 1$ numbers of the sequence. Observe that any consecutive $b$ of them can be expressed as an $x$-substring of the full string, with $a$ $y$-substrings removed; therefore, it has positive sum. On the other hand, any $(y - b)$-substring of the last $y + b - 1$ can be expressed as $a + 1$ $y$-strings with an $x$-string removed, hence has negative sum. It follows that $f(b, y - b) \geq y + b - 1$, but this contradicts our induction assumption because $b$ and $y - b$ are relatively prime.

To show that $f(x, y)$ is actually equal to $x + y - 2$ when $x$ and $y$ are relatively prime, we construct a string which has the required properties and more: It takes only two distinct values, and it is periodic with periods *both* $x$ and $y$. Call the two values $u$ and $v$, and imagine at first that we assign them arbitrarily as the first $y$ entries of our string.

Then these assignments are repeated until the end of the string, making the string perforce periodic in $y$. To be periodic in $x$ as well, we only need to ensure that the last $y - 2$ entries match up with the first $y - 2$, which entails satisfying $y - 2$ equalities among the original $y$ choices we made. Since there are not enough equalities to force all the choices to be the same, we can ensure that there is at least one $u$ and one $v$.

Let us do this, for example, with $x = 8$ and $y = 5$. Call the first five string entries $c_1, \ldots, c_5$, so the string itself will be $c_1 c_2 c_3 c_4 c_5 c_1 c_2 c_3 c_4 c_5 c_1$. To be periodic with period 8, we must have $c_4 = c_1$, $c_5 = c_2$, and $c_1 = c_3$. This allows us to have $c_1 = c_3 = c_4 = u$, for example, and $c_2 = c_5 = v$; the whole sequence is thus $uvuuvuvuuvu$.

Getting back to general $x$ and $y$, we note that a string which is periodic in $x$ automatically has the property that every $x$-substring has the same sum; because, as you shift the substring one step at a time, the entry picked up at one end is the same as the entry dropped at the other end. Of course the same applies to $y$-substrings if the string is periodic in $y$.

Let $S_x$ be the $x$-substring sum and $S_y$ similarly; we claim $S_x/x \neq S_y/y$. The reason is that if there are, say, $p$ copies of $u$ in each $x$-substring and $q$ copies of $v$ in each $y$-substring, then $S_x/x = S_y/y$ would imply $y(pu + (x-p)v) = x(qu + (y-q)v)$ which reduces to $yp = xq$. Since $x$ and $y$ are relatively prime, this cannot happen for $0 < p < x$ and $0 < y < q$.

It follows that we can adjust $u$ and $v$ so that $S_x$ is positive and $S_y$ is negative. In the case above, for example, each 8-substring contains 5 copies of $u$ and 3 of $v$, while each 5-substring contains 3 copies of $u$ and 2 of $v$. If we take $u = 5$ and $v = -8$, we get $S_x = 1$ and $S_y = -1$. The final sequence, solving the original problem, is then $5, -8, 5, 5, -8, 5, -8, 5, 5, -8, 5$. ♡

The industrious reader will not find it difficult to generalize the above arguments to the case where $x$ and $y$ have a greatest common divisor $\gcd(x, y)$ other than 1. The result is $f(x, y) = x + y - 1 - \gcd(x, y)$.

### First Odd Number in the Dictionary

This is just a matter of carefully and systematically considering the successive words involved in the description of a number. The earliest actual digit is of course "eight," but the earliest available word (or suffix for "eight") is "billion." Our number must begin with a digit so must start with "eight billion." Proceeding along these lines, one eventually gets the answer 8,018,018,885: "eight billion, eighteen million, eighteen thousand, eight hundred eighty-five." ♡

The idea for this silly puzzle came when Herb Wilf (University of Pennsylvania) asked me for the first *prime* in the dictionary. This question has been attributed to the computer guru Donald Knuth (Stanford University), and reasoning as above, followed by some checking on a computer, will lead you to 8,018,018,881.

# Combinatorics

Falsehood has an infinity of combinations, but truth has only one mode of being.

*---Jean-Jacques Rousseau*

If a puzzle begins with "How many ways are there to ...," it is automatically combinatorial, but the converse fails. Combinatorial reasoning is useful in the following (quite eclectic) list of puzzles and in many other puzzles in this book.

Our practice problem does fit the classical mold, however, and makes use of the most fundamental of combinatorial techniques: multiplying numbers of options.

## Sequencing the Digits

How many ways are there to write the numbers 0 through 9 in a row, such that each number other than the left-most is within one of some number to the left of it?

Solution: On the face of it, this problem does not seem amenable to multiplying numbers of options because the number of options depends on previous choices. For example, there are ten choices for the left-most digit, but if we start by writing "3" on the left, there are two choices for the next digit; if we start with "0" or "9," there is only one choice. If you know how to sum binomial coefficients, you can nonetheless analyze the problem in this manner, but there's a better way.

Observe that the sequence must terminate with a "0" or "9," and as we move *from right to left*, we always have a choice between writing the highest unused digit or the lowest—until we hit the left end, of course, where these two choices coincide.

Thus, there are two choices at each of nine opportunities. It follows that the total number of ways is $2^9 = 512$.  ♡

(*Source:* A Putnam Exam from the 1960s.)

21

The rest of the solutions are up to you. One hint: Keep your eyes open for more applications of the pigeon-hole principle!

## Subsets of Subsets

Prove that every set of ten distinct numbers between 1 and 100 contains two disjoint nonempty subsets with the same sum.

## The Malicious Maitre D'

At a mathematics conference banquet, 48 male mathematicians, none of them knowledgeable about table etiquette, find themselves assigned to a big circular table. On the table, between each pair of settings, is a coffee cup containing a cloth napkin. As each person is seated (by the maitre d'), he takes a napkin from his left or right; if both napkins are present, he chooses randomly (but the maitre d' doesn't get to see which one he chose).

In what order should the seats be filled to maximize the expected number of mathematicians who don't get napkins?

## Handshakes at a Party

Mike and Jenene go to a dinner party with four other couples; each person there shakes hands with everyone he or she doesn't know. Later, Mike does a survey and discovers that every one of the nine other attendees shook hands with a *different* number of people.

How many people did Jenene shake hands with?

## Three-Way Election

Ashford, Baxter, and Campbell run for secretary of their union, and finish in a three-way tie. To break it, they solicit the voters' second choices, but again there is a three-way tie. Ashford now steps forward and notes that, since the number of voters is odd, they can make two-way decisions; he proposes that the voters choose between Baxter and Campbell, and then the winner could face Ashford in a runoff.

Baxter complains that this is unfair because it gives Ashford a better chance to win than either of the other two candidates. Is Baxter right?

# The King's Salary

After the revolution, each of the 66 citizens of a certain country, including the king, has a salary of one dollar. The king can no longer vote, but he does retain the power to suggest changes—namely, redistribution of salaries. Each person's salary must be a whole number of dollars, and the salaries must sum to $n$. Each suggestion is voted on, and carried if there are more votes for than against. Each voter can be counted on to vote "yes" if his salary is to be increased, "no" if decreased, and otherwise not to bother voting.

The king is both selfish and clever. What is the maximum salary he can obtain for himself, and how long does it take him to get it?

# A Poorly Designed Clock

The hour and minute hands of a clock are indistinguishable. How many moments are there in a day when it is not possible to tell from this clock what time it is?

# A Mystifying Card Trick

David and Dorothy have devised a clever card trick. While David looks away, a stranger selects five cards from a bridge deck and hands them to Dorothy; she looks them over, pulls one out, and hands the remainder to David. David now correctly guesses the identity of the pulled card.

How do they do it? What's the biggest deck of cards they could use and still perform the trick reliably?

# Traveling Salesmen

Between every pair of major cities in Russia, there's a fixed air travel cost for going from either city to the other. Traveling salesman Alexei Frugal begins in Moscow and tours the cities, always choosing the cheapest flight to a city not yet visited (he does not need to return to Moscow). Salesman Boris Lavish also needs to visit every city, but he starts in Kaliningrad, and his policy is to choose the most *expensive* flight to an unvisited city at each step.

Prove that Lavish's tour costs at least as much as Frugal's.

## Losing at Dice

When six dice are rolled, the number of different numbers which can appear can range from 1 to 6. Suppose that once every minute, the croupier rolls six dice and you bet $1, at even odds, that the number of different numbers which appear will be exactly 4.

If you start with $10, roughly how long will it be, on average, before you are wiped out?

# Solutions and Comments

### Subsets of Subsets

The trick to this puzzle, based on a problem from the 1972 International Mathematical Olympiad, is to ignore the disjointness condition at first and just count subsets. A set $S$ of size 10 has, of course, $2^{10} - 1 = 1023$ nonempty subsets; can they all have different sums? The maximum sum of up to 10 numbers between 1 and 100 is $100 + 99 + \cdots + 91 < 1000$ and of course the minimum is 1, so by the pigeon-hole principle, there must be two distinct subsets $A \subset S$ and $B \subset S$ with the same sum.

Of course, $A$ and $B$ may not be disjoint, but you can just throw out their common elements; $A \setminus B$ (the set of elements of $A$ not in $B$) and $B \setminus A$ *are* disjoint and still have the same sum. ♡

## The Malicious Maitre D'

This problem can be traced to a particular event. Princeton mathematician John H. Conway came to Bell Labs on March 30, 2001 to give a "General Research Colloquium." At lunchtime, your author found himself sitting between Conway and computer scientist Rob Pike (now of Google), and the napkins and coffee cups were as described in the puzzle. Conway asked how many diners would be without napkins if they were seated in *random* order (see Chapter 11), and Pike said: "Here's an easier question—what's the *worst* order?"

If the maitre d' sees which napkin is grabbed each time he seats a diner (computer theorists would call him an "adaptive adversary"), it is not hard to see that his best strategy is as follows. If the first diner takes (say) his right napkin, the next is seated two spaces to his right so that the diner in between may be trapped. If the second diner also takes his right napkin, the maitre d' tries again by skipping another chair to the right. If the second diner takes his left napkin (leaving the space between him and the first diner napkinless), the third diner is seated directly to the second diner's right. Further diners are seated according to the same rule until the circle is closed, then the remaining diners (some of whom are doomed to be napkinless) are seated. This results in 1/6 of the diners without napkins, on average.

When, as in the stated puzzle, the maitre d' is *not* adaptive, it seemed likely to Pike and me (at the time) that the right strategy is to fill the even seats, then the odd ones. Each odd diner has probability 1/4 of finding himself napkinless, for an overall yield of 1/8 (6 of the 48 diners, on average).

However, a little more thought shows that the maitre d's best strategy is to use first the 0 mod 4 seats, then the odd seats, and finally the 2 mod 4 seats. This foils 9/64 of the diners on average, or $6\frac{3}{4}$ of the 48. To see this, call a diner "lonely" if when seated neither of his neighbors has yet appeared. We may assume all the lonely diners are seated first; note that there will be at most one napkinless diner between each consecutive pair of lonely diners.

Suppose two consecutive lonely diners are at distance $d$ (i.e., there are $d-1$ seats between them). Those seats will be filled in from either side; suppose the last diner between them finds himself at distances $a$ (to the right) and $b$ (to the left) from the lonely diners, where $a + b = d$. His right napkin will be gone unless the lonely diner to his right and all subsequent diners in between choose their left napkins; this occurs with probability $1/2^a$. Thus, the trapped diner loses (goes napkinless) with probability

$$(1 - 2^{-a})(1 - 2^{-b}) = 1 + 2^{-d} - 2^{-a} - 2^{-b} \, ,$$

which is minimized when $a$ and $b$ are equal or differ by 1.

If the lonely diners are spaced $d$ apart, we get one potential loser per $d$ diners, thus if the number $n$ of diners is a multiple of $d$, we find that the expected number of losers is $(n/d)(1-2^{-\lfloor d/2 \rfloor})(1-2^{-\lceil d/2 \rceil})$. It is easy to check that this quantity is maximized not at $d = 2$, where it is $n/8$, but at $d = 4$, where it is $9n/64$.  ♡

## Handshakes at a Party

This puzzle is an old chestnut and of the type that, at first glance, seems to offer insufficient information; why should we be able to deduce anything about *Jenene*? The answer is, ultimately, that Jenene is the partner of the one person not polled.

Since each person shook hands with at most eight others, the nine answers received by Mike are exactly the numbers 0 through 8. The two people (say, $A$ and $B$) who answered "0" and "8" must have been a couple, since otherwise their opportunity to shake one another's hand would have ruined one of those scores. Now we examine $C$ and $D$ who scored "1" and "7"; since $C$ had to shake hands with $A$ and $D$ had to miss $B$, the same argument applies and they must be a couple as well.

Similarly, the pairs scoring "2" and "6," and "3" and "5," must also be couples. This leaves both Mike and Jenene shaking hands exactly with the high scorers, for a score of "4" each.  ♡

If you didn't see the argument, but guessed that the answer was 4, your intuition is on track. In fact, if there is a unique answer (say, $x$), then $x$ must be 4, on account of symmetry. Suppose (for some reason) each couple had itself shaken hands, and Mike had asked everyone how many people had *not* shaken hands. Then the solution would have to be that Jenene shook $x+1$ hands. But switching the role of handshaking and nonhandshaking shows that $x + (x+1) = 9$.

Reasoning on the basis that the puzzle is a good one can be useful, if not entirely satisfactory. Martin Gardner, in one of his famous "Mathematical Games" columns in *Scientific American*, once asked: If a hole 6″ long is drilled through the center of a sphere, what is the remaining volume?

It seems like one would need to know the diameter of the hole, or of the original sphere, to solve the problem, but in fact, this is not the case. The bigger the sphere, the wider the hole has to be for its length to be 6″; a calculation verifies that the volume of the remaining napkin-ring-shaped solid is the same in every case.

However, you don't need to do that calculation if you trust the puzzle poser. The answer must be the same for a 6″ sphere with no hole at all, namely $\frac{4}{3}\pi 3^3 = 36\pi$ cubic inches.

## Three-Way Election

Baxter is correct—in fact, he has understated the case; assuming no voter changes his mind, Ashford will win for sure! To see this, suppose Ashford's supporters prefer Baxter to Campbell (so that Baxter would beat Campbell in the proposed two-candidate race). Then Baxter's supporters must prefer Campbell to Ashford, otherwise Campbell would have garnered fewer than 1/3 of the second-place votes; similarly Campbell's supporters prefer Ashford to Baxter. Thus, in this case, Ashford will beat Baxter in the runoff.

If Ashford's supporters prefer Campbell to Baxter, a symmetric argument shows that Ashford will beat Campbell in the runoff. ♡

This puzzle, dreamed up by mathematician Ehud Friedgut for classroom purposes, serves as a warning: There may be more to some tiebreakers than meets the eye!

## The King's Salary

This puzzle was devised by Johan Wästlund of Linköping University, and (loosely!) inspired by historical events in Sweden. There are two key observations: (1) that the king must temporarily give up his own salary to get things started, and (2) that the game is to reduce the number of salaried citizens at each stage.

The king begins by proposing that 33 citizens have their salaries doubled to $2, at the expense of the remaining 33 (himself included). Next, he increases the salaries of 17 of the 33 salaried voters (to $3 or $4) while reducing the remaining 16 to $0. In successive turns, the number of salaried voters falls to 9, 5, 3, and 2. Finally, the king bribes three paupers with $1 each to help him turn over the two big salaries to himself, thus finishing with a royal salary of $63.

It is not difficult to see that the king can do no better at any stage than to reduce the number of salaried voters to just over half the previous number; in particular, he can never achieve a unique salaried voter. Thus, he can do no better than $63 for himself, and the six rounds above are optimal. ♡

More generally, if the original number of citizens is $n$, the king can achieve a salary of $n-3$ dollars in $k$ rounds, where $k$ is the least integer greater than or equal to $\log_2(n-2)$.

## A Poorly Designed Clock

This delightful problem was posed by Andy Latto (andy.latto@pobox.com), a Boston-area software engineer, at the Gathering for Gardner IV, one of a series of conferences held in Atlanta in honor of Martin Gardner. It can be solved algebraically or geometrically, with sufficient care and patience, but there is an irresistible pencil-and-paperless proof, supplied to Andy by Michael Larsen, a mathematics professor at Indiana University. The idea of a third hand (instead of a second clock) came to me from David Gale.

Let us first note that for the problem to make sense, we must assume that the hands move continuously, and that we are not tasked with deciding whether a time is AM or PM. Note that we *can* tell what time it is when the two hands coincide, even though we can't tell which hand is which; this happens 22 times a day, since the minute hand goes around 24 times while the hour hand goes around twice, in the same direction.

This reasoning turns out to be good practice for the proof. Imagine that we add to our clock a third "fast" hand, which starts at 12 at midnight and runs exactly 12 times as fast as the minute hand.

Now we claim that whenever the hour hand and the fast hand coincide, the hour and minute hands are in an ambiguous position. Why? Because later, when the minute hand has traveled 12 times as far, it will be where the fast hand (and thus also the hour hand) is now, while the hour hand is where the minute hand is now. Conversely, by the same reasoning, all ambiguous positions occur when the hour hand and fast hand coincide.

So, we need only compute the number of times a day this co-incidence occurs. The fast hand goes around $12^2 \times 2 = 288$ times a day, while the hour goes around just twice, so this happens 286 times.

Of these 22 are times when the hour hand and minute hand (thus all three hands) are coincident, leaving 264 ambiguous moments.♡

### A Mystifying Card Trick

This card trick is usually credited to mathematician William Fitch Cheney. For more information, readers are referred to an article by Michael Kleber in the *Mathematical Intelligencer*, Vol. 24, No. 1 (Winter 2002); or Colm Mulcahy's article in the Mathematical Association of America's *Math Horizons*, in 2003, which discusses variations of the trick.

Dorothy communicates to David only through the ordering of the four cards she hands him. Of course, there are only $4! = 24$ orderings and seemingly 48 possibilities for the fifth card, but the key is that Dorothy gets to decide which of the original five cards is pulled.

The easiest way I know of to perform the trick is for Dorothy to pull a card in a suit that appears at least twice (the pigeon-hole principle again!). Suppose that suit is spades, and the cards $x$ and $y$ (thought of as numbers between Ace=1 and King=13, modulo 13). In one direction or the other, the cards must be at most 6 apart; let us assume $x$ is the "larger" so that $x - y \in \{1, 2, 3, 4, 5, 6\}$ mod 13. Thus, for example, we could have $x = 3 \equiv 16$ and $y = 12$ (Queen of spades) so that $x - y \equiv 4$.

Dorothy pulls $x$, puts $y$ first among the remaining four, and orders the three other cards so as to encode the difference $x - y$. For example, suppose David and Dorothy agree that the natural order of the deck is $\clubsuit A, \clubsuit 2, \ldots, \clubsuit K, \diamondsuit A, \ldots, \diamondsuit K, \heartsuit A, \ldots, \heartsuit K, \spadesuit A, \ldots, \spadesuit K$. If the three cards ascend (e.g., $\clubsuit 5, \clubsuit J, \diamondsuit 3$), then $x - y = 1$; call this the 123 order. We assign $x - y = 2$ to 132, $x - y = 3$ to 213, $x - y = 4$ to 231, $x - y = 5$ to 312, and finally $x - y = 6$ to 321.

It takes a little practice to do the trick smoothly.

Notice that there is some slack in the scheme; if fewer than four suits are represented among the five cards handed to Dorothy, she will have at least two choices for the pulled card. It is natural to ask how much bigger a deck could be accommodated; in fact, 124 cards is the maximum.

To see that you can do no better, imagine that the cards are numbered from 1 to $n$ and consider the function $f$ which assigns to any ordered 4-tuple $(u, v, y, z)$ with distinct entries the fifth card

$x$ which David is supposed to deduce from looking at the 4-tuple. For the trick to work, Dorothy must be able, given any set $S$ of five numbers in $\{1, \ldots, n\}$, to find a 4-tuple $(u, v, y, z)$ such that $S = \{u, v, y, z, f(u, v, y, z)\}$. Thus, the total number of 4-tuples must be at least equal to the total number of sets of size 5; i.e.,

$$n(n-1)(n-2)(n-3) \geq \binom{n}{5},$$

which implies $n-4 \geq 5!$, $n \geq 124$.

To actually accomplish the trick with cards numbered $1, \ldots, 124$ is surprisingly easy; here is a method suggested to me by Elwyn Berlekamp. Suppose the chosen cards are $c_1 < c_2 < \cdots < c_5$; Dorothy pulls card $c_j$, where $j$ is the sum of the values of all five cards, modulo 5. Looking at the remaining four, which sum (say) to $s$ modulo 5, David needs to find a number $x$ such that $x \equiv -s+k$ mod 5 if $x$ is $c_k$.

In other words, either $x$ is lower than any of David's cards, and satisfies $x \equiv -s+1 \mod 5$; or it is above the lowest card, but below the next one, and satisfies $x \equiv -s +2 \mod 5$; and so forth. But this is like saying that $x \equiv -s +1 \mod 5$ *if the remaining 120 cards are renumbered from 1 to 120* by closing the gaps left by David's four cards.

Since exactly $120/5 = 24 = 4!$ numbers from 1 to 120 have a given value modulo 5, we can neatly code the possibilities for $x$ by permuting David's four cards. ♡

## Traveling Salesmen

This puzzle, from the 11th All Soviet Union Mathematical Competition, Tallinn, 1977, is annoyingly tricky. *Obviously*, Lavish spends at least as much as Frugal! But how to prove it?

It seems that the best way is to show that for any $k$, the $k$th cheapest flight (call it $f$) taken by Lavish is at least as costly as the $k$th cheapest flight taken by Frugal. This seems like a stronger statement than what was requested, but it really isn't; if there was a counterexample, we could adjust the flight costs, without changing their order, in such a way that Lavish paid less than Frugal.

For convenience, imagine that Lavish ends up visiting the cities in west-to-east order. Let $F$ be the set of Lavish's $k$ cheapest flights, $X$ the departure cities for these flights, and $Y$ the arrival cities. Note that $X$ and $Y$ may overlap.

Call a flight "cheap" if its cost is no more than $f$'s; we want to show that Frugal takes at least $k$ cheap flights. Note that every flight eastward out of a city in $X$ is cheap, since otherwise it would have been taken by Lavish instead of the cheap flight in $F$ that he actually took.

Call a city "good" if Frugal leaves it on a cheap flight, "bad" otherwise. If all the cities in $X$ are good, we are done; Frugal's departures from those cities constitute $k$ cheap flights. Otherwise, let $x$ be the westernmost bad city in $X$; then when Frugal gets to $x$, he has already visited every city to the east of $x$, else Frugal could have departed $x$ cheaply. But then every city to the east of $x$, when visited by Frugal, had its cheap flight to $x$ available to leave on, so all are good. In particular, all cities in $Y$ east of $x$ are good, as well as all cities in $X$ west of $x$; that is $k$ good cities in all. $\heartsuit$

Thanks to Bruce Shepherd of Bell Labs for helping me come up with the above solution. We don't know what solution was intended by the composer.

## Losing at Dice

This is a trick, of course. On the average, it'll take forever for you to be wiped out—the game is in your favor! I noticed this counterintuitive fact years ago while constructing homework problems for an elementary probability course at Emory University.

There are $6^6 = 46,656$ ways to roll the dice. For four different numbers to appear, you need either the pattern AABBCD or AAABCD. There are

$$\binom{6}{2} \cdot \binom{4}{2}/2 = 45$$

versions of the former pattern, keeping the equinumerous labels alphabetical: e.g., AABBCD, ABABCD, ACDABB, but not BBAACD or AABBDC.

For the latter pattern, there are $\binom{6}{3} = 20$ versions.

In either case, there are $6 \cdot 5 \cdot 4 \cdot 3 = 360$ ways to assign numbers to the letters, for a total of $360 \cdot 65 = 23,400$ rolls. Thus, the probability of winning is $23400/46656 = 50.154321\%$. $\heartsuit$

If you win some bets with this game, don't forget to send 5% of your profits to me c/o A K Peters.

# Probability

> The human mind was designed by evolution to deal with foraging in small bands on the African savannah . . . faulting our minds for succumbing to games of chance is like complaining that our wrists are poorly designed for getting out of handcuffs.
>
> —*Steven Pinker, "How the Mind Works"*

Probability is with us every day. It forms the basis of the study of statistics, which in today's society plays a huge role in how decisions are made. But the historical origin of the theory of probability lies in games of chance and *gedanken*-experiments such as those you will see here.

Probability puzzles can be devastatingly counterintuitive. Consider the following reasonable-sounding question:

## Group Russian Roulette

In a room stand $n$ armed and angry people. At each chime of a clock, everyone simultaneously spins around and shoots a random other person. The persons shot fall dead and the survivors spin and shoot again at the next chime; eventually, either everyone is dead or there is a single survivor.

As $n$ grows, what is the limiting probability that there will be a survivor?

**Solution:** Amazingly, this probability does not tend to a limit; as $n$ grows, the probability varies subtly, but relentlessly, according to the fractional part of the natural logarithm of $n$. (For a related result, see H. Prodinger, "How to Select a Loser," *Discrete Math* **120** (1993) pp. 149–159.)

Our practice puzzle is honest, but closely related to the famous "Monty Hall Problem" (see below) which spawned a remarkable storm of confusion and controversy a decade ago.

# The Other Side of the Coin

A two-headed coin, a two-tailed coin and an ordinary coin are placed in a bag. One of the coins is drawn at random and flipped; it comes up "heads." What is the probability that there is a head on the other side of this coin?

Solution: Obviously, the coin chosen is either the fair coin or the two-headed coin, so its other side is equally likely to be a head or a tail, right? Wrong. You can think of it this way: If the coin were fair, it *might* have come up "tails," whereas the two-headed coin had no choice; hence, there is a presumption in favor of the two-headed coin. This notion is known to bridge players (and a century ago, to whist players) as the "principle of restricted choice."

To make it plainer, suppose the coin is tossed ten times and comes up "heads" every time. It *could* still be the fair coin, but we'd guess it was the two-headed coin. That presumption exists even after a single flip.

One way to calculate the odds in a straightforward manner is simply to think of the six *sides* of the coins as labeled: $H1$ and $H2$ on the two-headed coin, $T1$ and $T2$ on the two-tailed coin, and $H3$ and $T3$ on the fair coin. When a coin is drawn and flipped, each of the six sides is equally likely to appear. Of the three heads, $H1$ and $H2$ have a head on the other side, so the desired probability is $2/3$.♡

*Source:* Who knows. I used to perform this experiment myself when teaching elementary probability at Stanford and Emory Universities.

The Monty Hall Problem is based on the TV show, "Let's Make a Deal," on which (some) contestants were asked to choose one of three doors in search of a valuable prize. Host Monty Hall, who knew where the prize was, would open a second door instead: no prize there. The contestants were then given the option of sticking with their original choice or switching to the third door. I watched this show occasionally as a kid, and I remember audiences shouting to the contestant to "STAY!" or "SWITCH!" in about equal numbers.

Of course, she should switch. If the game is played 300 times, the right door will be chosen *initially* about 100 of those times; the other 200 games will be won by the contestant who switches!

If all this is obvious to you, do not despair. The remaining problems may yet test your confidence in your own probabilistic intuition.

# The Lost Boarding Pass

One hundred people line up to board an airplane, but the first has lost his boarding pass and takes a random seat instead. Each subsequent passenger takes his or her assigned seat if available, otherwise a random unoccupied seat.

What is the probability that the last passenger to board finds his seat occupied?

# Rolling All the Numbers

On average, how many times do you need to roll a die before all six different numbers have turned up?

# Odd Streak of Heads

On average, how many times do you need to flip a fair coin before you have seen a run of an odd number of heads, followed by a tail?

# Three Dice

You have an opportunity to bet $1 on a number between 1 and 6. Three dice are then rolled. If your number fails to appear, you lose $1. If it appears once, you win $1; if twice, $2; if three times, $3.

Is this bet in your favor, fair, or against the odds? Is there a way to determine this without doing any calculations?

# Magnetic Dollars

One million magnetic "susans" (Susan B. Anthony dollar coins) are tossed into two urns in the following fashion: The urns begin with one coin in each, then the remaining coins are thrown in the air one by one. If there are $x$ coins in one urn and $y$ in the other, then magnetism will cause the next coin to land in the first urn with probability $x/(x+y)$, and in the second with probability $y/(x+y)$.

How much should you be willing to pay, in advance, for the contents of the urn that ends up with fewer susans?

## Bidding in the Dark

You have the opportunity to make one bid on a widget whose value to its owner is, as far as you know, uniformly random between $0 and $100. What you do know is that you are so much better at operating the widget than he is, that its value to *you* is 80% greater than its value to him.

If you offer more than the widget is worth to the owner, he will sell it. But you only get one shot. How much should you bid?

## Random Intervals

The points $1, 2, \ldots, 1000$ on the number line are paired up at random, to form 500 intervals. What is the probability that among these intervals is one which intersects all the others?

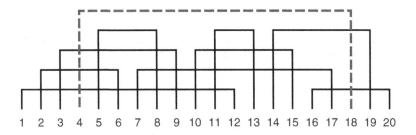

# Solutions and Comments

### The Lost Boarding Pass

We merely need to observe that when the 100th passenger finally boards, the seat remaining will be either the one assigned to him (or her) or the one assigned to the first passenger. Every other seat has been taken either by its rightful owner or by someone else who got there first.

Since there has been no preference exhibited at any stage toward one or the other of those two seats, the probability that the 100th passenger gets his own seat is 50%. ♡

The reasoning here is the same as that used, say, to compute the odds at craps. After you've rolled a "point" (4, 5, 6, 8, 9, or 10), you keep rolling until you get a 7 or another roll of your point. To compute the probability of winning (getting your point), you may assume that your next roll is the last one, and reckon accordingly. For example, if your point is 5, you win 4 times in 10 (because there are 4 ways to roll a 5 and 6 ways to roll a 7). In the case of the lost boarding pass, one of the first 99 passengers will eventually find his way either to $P_1$'s seat or to $P_{100}$'s seat, and here, given that one or the other was chosen, they are equally likely.

*Source:* Word-of-mouth. In this case, I heard the problem at the Gathering for Gardner V; Ander Holroyd supplied the version seen here.

## Rolling All the Numbers

This classic puzzle illustrates two important principles: mean waiting time and addition of expectations. Suppose you are repeating an experiment whose probability of success is $p$; how long do you have to wait, on average, to succeed? You can compute this value as a sum

$$\sum_{n=1}^{\infty} n(1-p)^{n-1}p = 1/p ,$$

but this is not very satisfactory from the point of view of intuition. Better is to imagine that the experiment is repeated $n$ times for $n$ so large that the fraction of successes is as close as you like to $p$ (law of large numbers). You can think of these $n$ trials as $pn$ separate series of experiments, each ending in a success; their average length is $n/(np) = 1/p$.

The puzzle calls for rolling all six numbers, and the key is to break up this process into six stages. The mean time it takes to complete all the stages is then the sum of the individual mean times of the stages. Here, you know that if you keep track of the *number* of different numbers you have seen, that this value begins at 1 (after the first roll) and rises a step at a time until it hits 6. We define "stage $k$" to be the period during which we have seen $k-1$ different numbers and are waiting to see the $k$th.

The probability of success during stage $k$ is just the number of numbers we haven't seen, namely $n - (k-1)$, divided by 6; thus, the mean length of stage $k$ is $6/(n-k+1)$. It follows that the mean time for the whole process is

$$\frac{6}{6} + \frac{6}{5} + \frac{6}{4} + \frac{6}{3} + \frac{6}{2} + \frac{6}{1} = 14.7 . \qquad \heartsuit$$

It is perhaps worth noting that it would be a quite different experiment to roll six dice at a time, waiting for all different numbers to appear on one such multiple roll. The probability of success there is $6 \cdot 5 \cdot 4 \cdot 3 \cdot 2 \cdot 1/6^6$ (see, e.g., the last problem of the previous chapter), which is about 0.0154321, so that the mean waiting time here is a huge 64.8 trials, even though you're rolling six dice in one trial!

## Odd Streak of Heads

This puzzle was suggested, but not used, for an IMO in the early '80s (see Murray Klamkin's *International Mathematical Olympiads 1979-1985*, Mathematical Association of America, 1986). It makes a nice mate to the previous puzzle, but requires a little more thinking.

If we compute the probability that we will start right out flipping an odd number of heads followed by a tail, we get $\Pr(HT) + \Pr(HHHT) + \Pr(HHHHHT) + \cdots = (\frac{1}{2})^2 + (\frac{1}{2})^4 + (\frac{1}{2})^6 + \cdots = \frac{1}{3}$. If we fail, it means we've hit a tail (after an even number of heads) and we have to start again. Thus, it will take three such experiments on average. But we want to count flips, not experiments.

Luckily, we can take advantage of another fact about expectations: If we have a random number $n$ of items whose average size is $s$, then the average total size of the items is $s$ times the average value of $n$. Each of our experiments (successful or not) ends when the first tail is flipped, so the average number of flips per experiment is $1/\frac{1}{2} = 2$. It follows that the solution to the puzzle is $2 \cdot 3 = 6$ flips.$\heartsuit$

However, there's another, nicer way to tackle this particular puzzle. Suppose $x$ is the answer. If we begin with T or HH, we still face an average of $x$ more flips before success; if we begin with HT, we've already succeeded. Thus,

$$ x = \frac{1}{2} \cdot (1 + x) + \frac{1}{4} \cdot (2 + x) + \frac{1}{4} \cdot 2 \ , $$

which gives us $x = 6$.

## Three Dice

This bet is, in fact, available in some casinos; in America it is called Chuck-a-Luck or Bird Cage (the dice are typically rolled in a cage). One might reasonably argue that this fact in itself is a calculation-free proof that the bet favors the house.

But there is a rather nice mathematical way to see this, which can be applied to other gambling games as well. Imagine that six players each bet $1 on a different number, then the dice are rolled. The house never loses! If three different numbers are rolled, the house just turns over the losers' $3 to the winners. Otherwise, the house collects $4 or $5 while paying out only $3.  ♡

So the game is in the house's favor if players bet this way, but does that mean it's *always* in the house's favor? Yes, it does—a bet favors the house or not independent of who bets and how much.

It is not, of course, difficult to determine directly that Chuck-a-Luck is a losing proposition. The probability of getting three different numbers from the dice is $6 \cdot 5 \cdot 4/6^3 = 5/9$, and the bettor breaks even on those since the probability of his number's being one of them is $1/2$. With probability $1/36$, all three dice have the same number; here the bettor wins $3 with probability $1/6$ and loses his $1 the rest of the time, for an average loss of $1/3$. Finally, the remaining $5/12$ of the time, the bettor wins $2 with probability $1/6$, wins $1 with probability $1/6$, and loses his $1 with probability $2/3$, for an average loss of $1/6$. Altogether this leaves him down by $1/36 \cdot 1/3 + 5/12 \cdot 1/6 = 17/216$ dollars, or about 8 cents per dollar bet.

The game can be made fair easily enough, by awarding the bettor $3 instead of $2 when he hits twice and $5 instead of $3 when he hits on all three dice.

This puzzle appeared in *Sam Loyd's Cyclopedia of 5000 Puzzles, Tricks, and Conundrums*, edited by Sam Loyd II, 1914. Sam Loyd (senior), 1841 to 1911, will be well known to most readers as a consummate showman and America's greatest puzzlist ever.

## Magnetic Dollars

Most people guess that the urn with fewer susans will be worth very little indeed, and in fact, at a restaurant table full of professional mathematicians recently, only one was willing to offer $100 and no one else would go higher than $10.

In fact, that urn is worth, on average, a cool quarter of a million dollars. The probability distribution of final contents for the two urns is exactly uniform: The probability that the first urn (say) will end up with just the one susan is the same as the probability that it will end up with 451,382 susans.

It is easy to prove this by induction, but I have found the following card-shuffling analogy to be more satisfying. Imagine a deck of 999,999 cards, just one of which is red. We will shuffle it perfectly as follows. Put the red card down on the table. Now take the next card (any card) and slip it with equal probability above or below the red card. The next card has three possible slots; choose one with equal probability and insert. When the last card is inserted, we have a perfectly random deck on the table.

But note: When there are $x-1$ cards above the red card, and $y-1$ below, the next card goes above with probability $x/(x+y)$. Thus, the cards above the red card function as susans (apart from the initial one) in the first urn, and the cards below, in the second urn.

Since in the final deck the red card is equally likely to be at any height, the uniformity of distribution for the susans follows. ♡

The puzzle (paradox?) of the Magnetic Dollars is sometimes called "Polya's Urn" after the late, great mathematician and puzzle enthusiast George Polya. (See, e.g., N. Johnson and S. Kotz, *Urn Models and Their Applications: An Approach to Modern Discrete Probability Theory*, Wiley, New York, 1977.) It is not hard to establish that if infinitely many susans are tossed, then with probability 1, the proportion of susans that fall in the first urn will approach a limit, drawn uniformly from the unit interval.

### Bidding in the Dark

You should not bid. If you do bid $\$x$, then the expected value to the widget's owner, *given that he sells*, is $\$x/2$; thus, its expected value to you, if you get it, is $1.8 \cdot \$x/2 = \$0.9x$. Thus you lose money on average if you win, and of course, you gain or lose nothing when you do not, so it is foolish to bet. ♡

*Source:* Maya Bar Hillel, University of Jerusalem.

### Random Intervals

This problem has a curious history. A colleague (Ed Scheinerman of Johns Hopkins University) and I needed to know the answer in order to compute the diameter of a random interval graph, and we at first computed an asymptotic value of 2/3. Later, using a lot of messy integrating, we found that the probability of finding an interval which intersects all others is *exactly* 2/3, for any number of intervals (from 2 on up).

The combinatorial proof below was found by Joyce Justicz, then taking a graduate reading course with me at Emory University. Suppose the interval endpoints are chosen from $\{1, 2, \ldots, 2n\}$. We will label the points $A(1)$, $B(1)$, $A(2)$, $B(2)$, $\ldots$, $A(n-2)$, $B(n-2)$ recursively as follows. Referring to points $\{n+1, \ldots, 2n\}$ as the *right side* and $\{1, \ldots, n\}$ as the *left side*, we begin by setting $A(1) = n$ and letting $B(1)$ be its mate. Suppose we have assigned labels up to $A(j)$ and $B(j)$, where $B(j)$ is on the left side; then $A(j+1)$ is taken as the left-most point on the right side not yet labeled, and $B(j+1)$ as its mate. If $B(j)$ is on the right side, $A(j+1)$ is the right-most unlabeled point on the left side and again $B(j+1)$ is its mate.

If $A(j) < B(j)$, we say that the $j$th interval "went right," otherwise it "went left." Points labeled $A(\cdot)$ are said to be *inner* endpoints, the others *outer*.

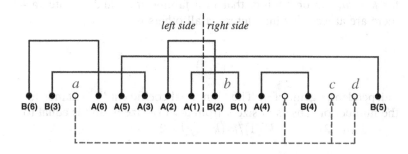

It is easily checked by induction that after the labels $A(j)$ and $B(j)$ have been assigned, either an equal number of points have been labeled on each side (in case $A(j) < B(j)$) or two more points have been labeled on the left (in case $A(j) > B(j)$).

When the labels $A(n-2)$ and $B(n-2)$ have been assigned, four unlabeled endpoints remain, say $a < b < c < d$. Of the three equiprobable ways of pairing them up, we claim two of them result in a "big" interval which intersects all others, and the third does not.

In case $A(n-2) < B(n-2)$, we have $a$ and $b$ on the left and $c$ and $d$ on the right, else only $a$ is on the left. In either case, all inner endpoints lie between $a$ and $c$, else one of them would have been labeled. It follows that the interval $[a, c]$ meets all others, and likewise $[a, d]$, so unless $a$ is paired with $b$, we get a big interval.

Suppose on the other hand that the pairing is indeed $[a, b]$ and $[c, d]$. Neither of these can qualify as a big interval since they do not intersect each other; suppose some other interval qualifies, say $[e, f]$, labeled by $A(j)$ and $B(j)$.

When $a$ and $b$ are on the left, the inner endpoint $A(j)$ lies between $b$ and $c$, thus $[e, f]$ cannot intersect both $[a, b]$ and $[c, d]$, contradicting our assumption.

In the opposite case, since $[e, f]$ meets $[c, d]$, $f$ is an outer endpoint (so $f = B(j)$) and $[e, f]$ went right; since the last labeled pair went left, there is some $k > j$ for which $[A(k), B(k)]$ went left, but $[A(k\text{--}1), B(k\text{--}1)]$ went right. Then $A(k) < n$, but $A(k) < A(j)$ since $A(k)$ is a later-labeled, left-side inner point. But then $[A(j), B(j)]$ does not, after all, intersect $[B(k), A(k)]$, and this final contradiction proves the result. ♡

With slightly more care one can use this argument to show that for $k < n$, the probability that in a family of $n$ random intervals there are at least $k$ which intersect all others is

$$\frac{2^k}{\binom{2k+1}{k}}$$

independent, again, of $n$. The "binomial coefficient" $\binom{n}{k}$ stands for the number of subsets of size $k$ from a set of size $n$, and is equal to $n(n-1)(n-2) \cdots (n-k+1)/k((k-1)(k-2) \cdots 1$.

# Geometry

> Equations are just the boring part of mathematics. I attempt to see
> things in terms of geometry.
>
> —*Stephen Hawking (1942–)*

Classical geometry in two or three dimensions is a bottomless well
for problem composers, but to be a *puzzle*, we ask that it not be
something Euclid would have included in his Volume II. Hence, you
won't be asked here to show AB=CD or this triangle is congruent
to that one.

Fortunately there is still a great variety of fascinating geometry
puzzles from which to choose.

Our practice problem showed up in 1980 on the Preliminary
Scholastic Aptitude Exam, but, to the embarrassment of the Ed-
ucational Testing Service, the answer they marked as correct was
wrong. A confident student called the ETS to task when his exam
was returned to him. Luckily for us, the *correct* answer boasts a
marvelous, intuitive proof. (*Note:* Subsequently the ETS created a
panel, on which your author served, for reviewing the questions on
their mathematics aptitude tests.)

## Gluing Pyramids

A solid square-base pyramid, with all edges of unit length, and a
solid triangle-base pyramid (tetrahedron), also with all edges of unit
length, are glued together by matching two triangular faces.

How many faces does the resulting solid have?

**Solution:** The square-base pyramid has five faces and the tetrahedron four. Since the two glued triangular faces disappear, the
resulting solid has $7 = 5+4-2$ faces—right? This, apparently, was
the intended line of reasoning. It may have occurred to the composer that in theory, some pair of faces, one from each pyramid,
could in the gluing process become adjacent and coplanar. They
would thus become a single face and further reduce the count. But,

43

*surely*, such a coincidence can be ruled out. After all, the two solids are not even the same shape.

In fact, this does happen (twice): The glued polyhedron has only five faces.

You can see this in your own head. Imagine *two* square-based pyramids, sitting side-by-side on a table with their square faces down and abutting. Now, draw a mental line between the two apexes; observe that its length is one unit, the same as the lengths of all the pyramid edges.

Thus, between the two square-based pyramids, we have in effect constructed a regular tetrahedron. The two planes, each of which contains a triangular face from each square-based pyramid, also contain a side of the tetrahedron; the result follows.     ♡

(Check the figure below if you find this hard to visualize.)

This argument, sometimes called the "pup tent" solution, appeared in a 1982 article by Steven Young, "The Mental Representation of Geometrical Knowledge," in the *Journal of Mathematical Behavior*.

One of the puzzles below has a "no-word proof"—a picture suffices. See if you can guess which one.

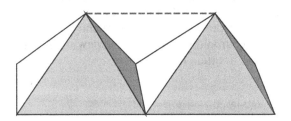

## Circles in Space

Can 3-space be partitioned into circles?

## Magic with Cubes

Can you pass a cube through a hole in a smaller cube?

# Red Points and Blue Points

Given $n$ red points and $n$ blue points on the plane, no three on a line, prove that there is a matching between them so that line segments from each red point to its corresponding blue point do not cross.

# Line through Two Points

Suppose $X$ is a finite set of points on the plane, not all on one line. Prove that there is a line passing through exactly two points of $X$.

# Pairs at Maximum Distance

Again, $X$ is a finite set of points on the plane. Suppose $X$ contains $n$ points and the maximum distance between any two of them is $d$. Prove that at most $n$ pairs of points of $X$ are at distance $d$.

# Monk on a Mountain

A monk begins an ascent of Mt. Fuji on Monday morning, reaching the summit by nightfall. He spends the night at the summit and starts down the mountain the following morning, reaching the bottom by dusk on Tuesday.

Prove that at some precise time of day, the monk was at exactly the same altitude on Tuesday as he was on Monday.

# Painting the Polyhedron

Let $P$ be a polytope with red and green faces such that every red face is surrounded by green ones, but the total red area exceeds the total green area. Prove that you can't inscribe a sphere in $P$.

# Circular Shadows

The projections of a solid body onto two planes are perfect disks. Prove that they have the same radius.

# Strips in the Plane

A "strip" is the region between two parallel lines on the plane. Prove that you cannot cover all of the plane with a set of strips the sum of whose widths is finite.

# Diamonds in a Hexagon

A large regular hexagon is cut out of a triangular grid and tiled with diamonds (pairs of triangles glued together along an edge). Diamonds come in three varieties, depending on orientation; prove that precisely the same number of each variety must appear in the tiling.

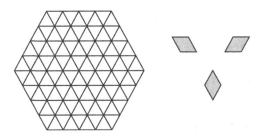

# Rhombus Tiling

Let's do this one again, but with bigger tiles and more sides.

Form $\binom{n}{2}$ different rhombi from the pairs of nonparallel sides of a regular $2n$-gon, then tile the $2n$-gon with translations of the rhombi. Prove you use each different rhombus exactly once!

# Vectors on a Polyhedron

To each face of a polytope we associate an out-pointing vector perpendicular to that face, with length equal to the area of the face. Prove that the sum of these vectors is zero.

# Three Circles

The "focus" of two circles is the intersection of two lines, each of which is tangent to both circles, but does not pass between them.

Thus three circles of different radii (but none contained in another) determine three foci. Prove that the three foci lie on a line.

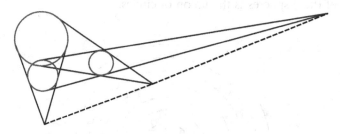

## Sphere and Quadrilateral

A quadrilateral in space has all of its edges tangent to a sphere. Prove that the four points of tangency lie on a plane.

The last puzzle is an excursion into topology, and different sizes of infinity.

## Figure 8s in the Plane

How many disjoint topological "figure 8s" can be drawn on the plane?

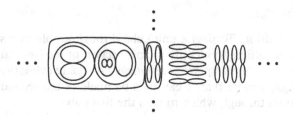

## Solutions and Comments

### Circles in Space

Yes. Place a circle of radius 1 on the $XY$ plane, centered at each 1 mod 4 point on the $X$-axis (that is, ..., (-7,0), (-3,0), (1,0),

(5,0), (9,0) ...). Observe that every sphere centered at the origin hits the union of these circles at precisely two points. The rest of each of these spheres is the union of circles.    ♡

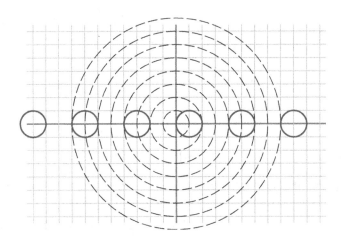

There are other ways to do this, e.g., involving tori, but nothing I know of approaches the above in elegance and simplicity.

I first heard this cute dissection puzzle from Nick Pippenger, Professor of Computer Science at Princeton University.

## Magic with Cubes

You can do it. To pass a unit cube through a hole in a second unit cube, it suffices to identify a cross section of the (second) cube which contains a unit square *in its interior*. A square cylindrical hole of side slightly more than 1 can then be made in the second cube, leaving room through which to pass the first cube.

You can then do the same, with even smaller tolerances, if the second cube is just a bit smaller than a unit cube.

The easiest (but not the only) cross section you can try this with is the regular hexagon you get by slicing through three vertices and the centroid. You can see this hexagon by viewing the cube so that one of its vertices is in the center.

Letting $A$ be the projection of one of the visible faces onto the plane, we observe that its long diagonal is the same length ($\sqrt{2}$) as a unit square's, since that line has not been foreshortened. If we slide

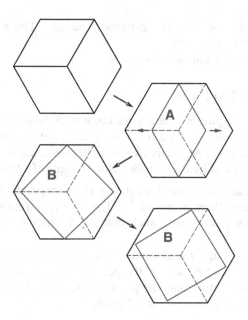

a copy of $A$ over to the center of the hexagon, then widen it to form a unit square $B$, $B$'s widened corners will not reach the vertices of the hexagon (since the distance between opposing vertices of the hexagon exceeds the distance between opposing sides).

It follows that if we now tilt $B$ slightly, all four of its corners will lie strictly inside the hexagon.                                       ♡

I was reminded of this charming puzzle, which has appeared in a Martin Gardner column, by Gregory Galperin of Eastern Illinois University.

## Red Points and Blue Points

Among all matchings, take one which minimizes the total length of the $n$ connecting line segments; we claim this cannot have any crossings. For, if the segment $uv$ crosses $xy$, then these two segments are the diagonals of the convex quadrilateral $uyxv$ and, using the triangle inequality, we see that using the sides $uy$ and $xv$ would have reduced the total length.                                       ♡

The general technique used here, finding a object with specific properties by looking for something that minimizes or maximizes some parameter, is sometimes called the *variational method* and is, as many readers will know, extremely useful. The next puzzle provides another example.

*Source:* A Putnam Exam from the 1960s.

### Line through Two Points

This famous puzzle was a conjecture of Sylvester dating back to 1893. It was first proved by Tibor Gallai, but the proof below, found in 1948 by L. M. Kelly (*American Mathematical Monthly*, Vol. 55), was often cited by Paul Erdös as an example of a "book proof."

Assume that every line through two or more points in $X$ in fact contains at least three points of $X$. The idea is to find such a line $L$, and a point $P$ not on $L$, such that the distance from $P$ to $L$ is minimized.

Since $L$ contains at least three points of $X$, two of them, say $Q$ and $R$, lie on the same side of the perpendicular to $L$ from $P$. But then, if $R$ is the farther one, the point $Q$ is closer to the line through $P$ and $R$ than $P$ is to $L$—contradiction. ♡

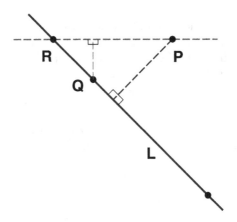

### Pairs at Maximum Distance

To solve this puzzle, from the 1957 Putnam Exam, it's useful to observe that if $A, B$ and $C, D$ are two "max pairs" (pairs of points from $X$ at distance $d$), then the line segments $AB$ and $CD$ must cross (else one of the diagonals of the quadrilateral $ABDC$ would exceed $d$ in length).

Geometry

Now assume the statement of the puzzle is false, and let $n$ be the size of a smallest counterexample. Since there are more than $n$ max pairs and each has two points, there must be a point $P$ which participates in three max pairs (say, with points $A$, $B$, and $C$). Every two of the segments $PA$, $PB$, and $PC$ must make at most a 60° angle at $P$, and one of them, say $B$, must lie between the others.

But this makes it pretty tough for $B$ to be in any other max pair, since if $BQ$ were a max pair, it would have to intersect both $PA$ and $PC$—an impossibility. Thus we can drop $B$ out of $X$ altogether, losing only one max pair and obtaining a smaller counterexample. This contradiction completes the proof. ♡

Monk on a Mountain

Perhaps the easiest way to see this is to imagine that the monk has a twin who is instructed to climb the mountain on Tuesday exactly as the monk himself climbed it on Monday. The monk must pass his twin on the way down on Tuesday, or, if they are not on the same path, must at some point be at the same altitude as his twin.♡

(Perhaps you found this puzzle too easy; fear not, a much more challenging version awaits you in Chapter 11—Toughies.)

This ancient puzzle can be viewed as an application of the very useful Intermediate Value Theorem, which says that a continuous function must pass through all intermediate values. The function in this case can be taken to be the difference between the monk's altitude at a particular time of day on Monday, and the same time on Tuesday; that function begins negative (at about minus the height of Mt. Fuji) and ends positive, so must at some point have been zero.

Geometrically speaking, you can imagine that the monk's altitude for each day is plotted on a graph, and the two graphs are superimposed. There must be a place (or places) where they cross.

Other famous applications of the Intermediate Value Theorem include inscribing Lake Michigan in a square, and cutting a ham sandwich (with a planar slice) so that the bread, ham, and cheese are all exactly bisected.

## Painting the Polyhedron

Assume the sphere is inscribed, and triangulate the faces of $P$ using the points where the sphere is tangent. Then the triangles on either side of any edge of $P$ are congruent, thus have the same area; at most, one of every such pair of triangles is red. It follows that the red area is at most equal to the green area, contradicting our assumptions. ♡

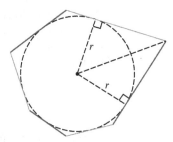

This puzzle came my way from Emina Soljanin of Bell Labs. The illustration shows a two-dimensional version, where sides and vertices of a polygon take the place of faces and edges of $P$.

## Circular Shadows

This potentially frustrating conundrum comes from the 5th All Soviet Union Mathematical Competition, Riga, 1971. An easy way to make your intuition rigorous is to select a plane which is simultaneously perpendicular to the two projection planes, and move parallel copies of it toward the body from each side. They hit the body at the opposite edges of each projection, so that the distance between the parallel planes at that moment is the common diameter of the two projected circles. ♡

## Strips in the Plane

Like the previous problem, this one, a version of which appeared on an early Putnam Exam, presents another "intuitively obvious" fact for you to prove.

Since it's hard to compare infinite volumes, it makes sense to focus on some finite part of the plane. We can't control the relative angles of the strips, so it is logical to look at a disk $D$ of radius $r$.

Assume the strips have widths $w_1$, $w_2$, ... which sum to 1; it turns out that they can't even cover $D$ in the $r = 1$ case. The intersection of $D$ with strip of width $w$ is contained in a rectangle of width $w$ and length 2, and therefore has area less than $2w$. Thus, the total area inside $D$ covered by the strips is less than 2, but the area of $D$ is of course $\pi > 2$. ♡

This argument shows that you need the widths of the strips to sum to more than $\pi/2$ to cover the unit disk, but in fact you can't do it unless the sum is at least 2 (in which case parallel strips will do the job). There is quite a lovely proof of this fact: The idea is to extend the puzzle into 3-space by taking $D$ to be a cross section through the center of a unit ball. Suppose the disk is covered by strips of total width $W$, and let $S$ be one of the strips, of width, say, $w$. We may assume that either both edges of the strip cross $D$ or one crosses and the other is tangent. Projecting $S$ upward and downward to the surface of the ball, we get a belt (or cap) encircling the ball—whose area, one can show using calculus, is $2\pi w$ independent of the position of the strip!

Since the total surface area of the ball is $4\pi$, you need $W \geq 2$ to cover it, and if you aren't covering the surface of the ball, you aren't covering the disk.

## Diamonds in a Hexagon

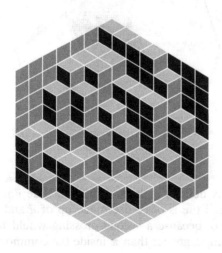

♡

Proofs without words have become a regular feature in two Mathematical Association of America journals, *Mathematics Magazine* and *The College Mathematics Journal*. You can can find these reprinted in the books, *Proofs Without Words* and *Proofs Without Words II*, by Roger B. Nelsen, published by the MAA. Diamonds in a Hexagon appears in the first volume as "The Problem of the Calissons."

## Rhombus Tiling

Let $\vec{u}$ be one of the sides of the $2n$-gon; a $\vec{u}$-rhombus is any of the $n-1$ rhombi using $\vec{u}$ as one of its two vectors. In a tiling, the tile next to a $\vec{u}$-side must be a $\vec{u}$-rhombus, as must the tile on the other side of that one, and so forth until we reach the opposite side of the $2n$-gon. Notice that each step of this path proceeds in the same direction (that is, right or left) with respect to the vector $\vec{u}$, as must any other path of $\vec{u}$-rhombi; but then there can be no other $\vec{u}$-rhombi, since they would generate paths with no way to close and nowhere to go.

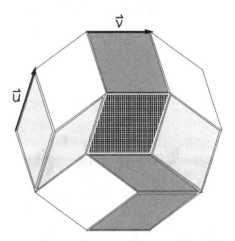

The similarly defined path for a different side $\vec{v}$ must cross the $\vec{u}$, and the shared tile is of course made up of $\vec{u}$ and $\vec{v}$. Can they cross twice? No, because a second crossing would have $\vec{u}$ and $\vec{v}$ meeting at an angle greater than $\pi$ inside the common rhombus. ♡

This puzzle reached me from Dana Randall of Georgia Tech.

# Geometry

## Vectors on a Polyhedron

This puzzle was brought to my attention by Yuval Peres of the Department of Statistics at UC Berkeley. The easiest way to see that the vectors must sum to zero is to perform the following *gedanken-experiment*: Pump air into the (rigid) polytope, and observe that the pressure on a face is a force acting in the direction of the normal and of magnitude proportional to the area. These pressures must balance, otherwise the polytope would move of its own accord! ♡

## Three Circles

This is the best example I know of the effectiveness of moving a puzzle *up* in dimension. Replace each circle by a sphere whose intersection with the plane is the given circle; now each pair of spheres determines a cone, and the apexes of these three cones are the three points in question.

But these apexes all lie on the plane which is tangent to the spheres from above; and similarly, on the plane tangent to the spheres from below. Hence, they lie on the intersection of these two planes: a line! ♡

This seems to be an ancient, classical puzzle. I heard it first from Dana Randall, of the College of Computing at Georgia Tech. Vadim Zharnitsky, of the University of Illinois, has observed that you can ask a similar question about four spheres in 3-space: Do the apexes of the six cones they determine lie on a plane? In fact, they do, and one way to prove it is to lift once more, to the fourth dimension.

## Sphere and Quadrilateral

This puzzle was obtained from Tanya Khovanova, a Visiting Research Staff Member in the Program in Applied and Computational Mathematics at Princeton University. She keeps a list of what she calls "coffin" problems. In her words:

> The Mathematics Department of Moscow State University, the most prestigious mathematics school in Russia, had at that time [1975] been actively trying to keep Jewish students (and other "undesirables") from being able to enroll at the Department. One of the methods they used for doing this was giving the unwanted students a different set of problems on the oral exam. These problems were carefully designed to have

an elementary solution (so that the Department could avoid scandals) that was nearly impossible to find. Any student that failed to answer could be easily rejected, so this system was an effective method of controlling admissions. These kinds of problems were informally referred to as "coffins."

The following solution would indeed be tough to find, but perhaps not impossible if you appreciate that a nice way to prove four points are coplanar is to find a point which is on lines through disjoint pairs of the four points. Start by observing that each vertex $i$ of the quadrilateral is at the same distance $d_i$ from each of the two points of tangency of its incident edges. Attribute a mass $1/d_i$ to vertex $i$; then the center of mass for any two adjacent vertices is the point of tangency of their common edge. It follows that the center of mass of all four vertices lies on the line connecting opposite points of tangency, so this point does the trick. ♡

## Figure 8s in the Plane

This puzzle has been around for 50 years or so; I once heard it attributed to the late, great topologist R. L. Moore of the University of Texas. Readers unfamiliar with differing degrees of "infinity" will already have been confused; clearly, you can draw infinitely many disjoint figure 8s on the plane, for instance, by placing a small one inside each box of a square grid. Such a collection is said to be "countable," meaning that one could number the 8s with positive integers in such a way that each 8 gets a different number.

The set of all integers, the set of all *pairs* of integers, and thus the set of all rational numbers, are all countable, but as the brilliant (but frequently depressed) mathematician Georg Cantor observed in 1878, the set of all *real* numbers is *not* countable. We could draw concentric circles on the plane with all possible positive real diameters, hence if the puzzle asked for circles instead of figure 8s, the answer would be "uncountably many," or more precisely, "the cardinality of the reals."

However, we can only draw countably many 8s. Associate with each 8 a pair of rational points (points of the plane with both coordinates rational numbers), one in each loop; no two figure 8s can share a pair of points. Hence, the cardinality of our set of 8s is no greater than the set of pairs of pairs of rational numbers, which is countable. ♡

See Chapter 11 (Toughies) for a trickier version of this puzzle.

# Geography(!)

Without geography you're nowhere.
—*Jimmy Buffett (1946–)*

OK, this chapter does not belong in the book. Some of the puzzles are mathematical in nature, to be sure, but really they are here because mathematical puzzle lovers seem to enjoy them. My publisher has assured me that the book's price would be the same without this chapter, so it's free and you can skip it with a clear conscience.

The basis for the puzzles is the surface of planet Earth. There is a heavy bias, however, toward my own country, the United States; readers from other countries, please forgive me. I will be grateful for similar puzzles, related to other countries, sent to me at pw@akpeters.com.

Several of the puzzles test the degree to which planar projections have distorted our understanding of the globe. Here's a sample:

## Out of Africa

Which is the closest US state to Africa?

**Solution:** Maine. ♡

It's not even close; check a globe. If you fly the great circle from Miami to (say) Casablanca, you'll start off going NE up the eastern seaboard, not missing Maine by much.

You're on your own now.

## East of Reno

What's the biggest city in the US east of Reno, Nevada and west of Denver, Colorado?

# The Phone Call

A phone call is made from an East Coast state to a West Coast state, and it's the same time of day at both ends. How can this be?

# The Diameter of the US

Which two states contain the most distant pair of points in the United States?

# South from Key West

If you fly due south out of Key West, Florida, which South American country will you hit first?

# Indians in the Midwest

Which is the only Midwestern US state whose name is not of Native American origin?

# The Largest Second-Largest City

Which is the largest city in the US which is not the largest city in the US of its name?

OK, maybe that's a little confusing. Let's put it another way: Say that a (US) city is "eclipsed" if there's a larger one with the same name; for example, Portland, Maine is eclipsed (by Portland, Oregon). What we're asking for is the largest eclipsed city.

# The Natural Border

Some parts of state borders are natural (determined by bodies of water, mountains, etc.) while others are legislated lines—in one famous case (involving Delaware and Pennsylvania), an arc of a circle. Three states—Colorado, Utah, and Wyoming—have only artificial borders. Which state has only natural borders?

# The Uncrossable Border

Speaking of state borders, can you find one which cannot be crossed by car? In other words, find two states which share a border, yet you cannot pass directly from one to the other in your automobile.

# Department of Odd Names

What distinction is held by the point of land called West Quoddy Head, Maine?

# Urban and Rural

Here's a more sociological puzzle. These days most Americans—over 75%—live in a "metropolitan area." The 2000 census lists 100% of the population of one state as metropolitan, but only 27.6% (the least) of another only a few hundred miles away. Can you guess these two states?

# Cities North and South

How's your visualization of the continents? Put these four cities in order, South to North: Halifax, Nova Scotia; Tokyo, Japan; Venice, Italy; Algiers, Algeria.

# The One-Syllable City

What's the largest city in the U.S. with a one-syllable name?

# Washingtons and Feminists

Here's a test of your mental map of the continental US. Can you design a automobile trip in the US, beginning in Seattle, Washington and ending in Washington, DC, which enters only states whose names begin with one of the letters in the word WOMAN?

Our final geography puzzle signals a (slight) return to mathematics.

Mathematical Puzzles

## The Naturalist and the Bear

A naturalist left her camp, hiked 10 miles south, and then 10 miles east, where she spotted and photographed a bear. She then hiked 10 miles north and was back at her camp.

You haven't got the photograph, but you still know what color the bear was, right?

# Solutions and Comments

Solutions to these puzzles can be verified with an atlas, globe, almanac, or the 2000 US Census Report. Let's see how well you guessed...

### East of Reno

"Biggest city" questions can be awkward; the standard is to measure by population (not area!) within the legal city limits, which of course can be misleading with respect to metropolitan areas. For example, the almanac figures make Jacksonville, Florida appear to be larger than Atlanta, Georgia even though metro area population of the latter is almost four times greater.

However, there is no such subtlety needed here. The biggest city east of Reno and west of Denver is, by any measure, Los Angeles, California.

♡

### The Phone Call

"East Coast" states run from Maine down to Florida. "West Coast" certainly includes Washington, Oregon and California, to which you can add Alaska and even Hawaii if you wish—but it doesn't help.

Within the continental US, such a phone call would normally face a three-hour time difference. We can eliminate one of those hours by calling from the western portion of the Florida panhandle (say, Pensacola), which lies in the Central Time Zone; and another by calling from one of several towns in far eastern Oregon (say, Ontario) which observes Mountain Time.

The last hour disappears if we carefully call between 2 and 3 hours after midnight in Pensacola, on the morning when Daylight

Saving Time ends in late October. Central Time will have been retarded by an hour at that point, but Mountain Time not yet affected.
♡

## The Diameter of the US

Obviously, it's either Hawaii and Maine, or Alaska and Florida. Or is it Hawaii and Alaska?

Amazingly, it's none of the above: The correct answer is Hawaii and Florida. The great circle strikes again!   ♡

## South from Key West

This is, of course, a trick question. You won't hit any country in South America; you will pass *west* of the entire continent.   ♡

## Indians in the Midwest

With a liberal definition of "Midwestern state" you might include Minnesota, Wisconsin, Iowa, Illinois, Missouri, Michigan, Ohio, Kansas, and Nebraska—all with names of Native American origin—and the answer, Indiana!   ♡

Curiously, only one state east of the Mississippi has a *capital* whose name is of Native American origin—Florida (Tallahassee).

## The Largest Second-Largest City

Portland, Maine? Springfield, Something? Popular guesses, but not correct. Before 1975 or so, the right answer would have been Kansas City, Kansas, eclipsed of course by Kansas City, Missouri. Then for a while the winner was Columbus, Georgia, eclipsed by the capital of Ohio. However, we are in the age of suburbia, and the 2000 census shows that Glendale, California (eclipsed by Glendale, Arizona) now owns this obscure honor.   ♡

## The Natural Border

Hawaii, of course, has all natural borders. Perhaps you thought this was too easy, but folks often have a blind spot.   ♡

## The Uncrossable Border

Much harder. Wisconsin and Michigan share a long border in Lake Michigan, but you can cross it sitting in your car in the Manitowoc–Ludington Ferry. There's a ferry from Montauk Point (NY) to Block Island (RI) which crosses the not-very-well-known border between those two states, and this one is passengers-only. Other solutions are also possible. ♡

A related question asks for a *piece* of a state which is accessible by car from the rest of the state only by passing through another state (or Canada, as in the case of Point Roberts, WA). There are a number of such places, especially near the ever-changing Mississippi River.

## Department of Odd Names

West Quoddy Head is the *easternmost* point of the continental US. ♡

You will sometimes hear it said that Cape Wrangell, on Attu Island in Alaska, is the winner if you drop the word "continental," but I don't buy that Greenwich-centered reasoning. Would you call Cape Wrangell the easternmost point *of Alaska*?

## Urban and Rural

New Jersey and Vermont. For this and lots of other interesting tidbits, you might want to check http://www.census.gov/prod/2002pubs/01statab/pop.pdf. ♡

## Cities North and South

Tokyo, Algiers, Halifax and finally Venice. The latitudes are, respectively, 35°40′N; 36°50′N; 44°53′N; and 45°26′N. Notice that the 45th parallel separates the last two, making it easy to see that Venice is the more northerly. I won $1 from a Nova Scotian with this one once. ♡

## The One-Syllable City

York, Pennsylvania and Troy, New York are frequent (and good) guesses but Flint, Michigan, despite substantial loss of population, remains the only one-syllable city in the US with a population of more than 100,000. However, if you go by local pronunciation, the winner is, arguably, Newark ("Noork"), New Jersey! ♡

## Washingtons and Feminists

No problem. Drive South through Oregon, Nevada and Arizona, east through New Mexico, into the Oklahoma panhandle, and out the NE corner of Oklahoma into Missouri. Here, you must turn north and exit the NW corner of the state into Nebraska, continuing west into Wyoming and north into Montana—a long trip just to avoid Idaho. Finally, you can turn east again through North Dakota, Minnesota, Wisconsin, and Michigan. Now south into Ohio and east through West Virginia into Maryland, and on into Washington, DC. ♡

You'll have to leave the interstate highway system several times (or for a long period) to actually drive this tour, but presumably you're not in a rush.

## The Naturalist and the Bear

The original idea was, of course, that the naturalist's camp must have been at the North Pole for her path (10 miles south, 10 miles east, then 10 miles north) to have been a closed loop; thus, the bear was white (a polar bear). However, as pointed out in one of Martin Gardner's columns, there are infinitely many other points on the globe where such a path would close.

Some of these points lie on a circle slightly less than $10 + 5/\pi$ miles in radius about the South Pole, from which the first leg of the hike would take the naturalist to a point $P$ a hair less than $5/\pi$ miles from the South Pole. The trek 10 miles east will carry her all the way around the world and back to $P$, from which her northward leg will of course take her back to camp.

Another circle a bit less than $10 + 5/2\pi$ miles in radius will also work, the naturalist encircling the pole twice during her eastern leg, and so on.

There are no bears in Antarctica, and if there were, they would probably be white, so the answer to the puzzle is unchanged. ♡

# Games

Money was never a big motivation for me, except as a way to keep
score. The real excitement is playing the game.
    —*Donald Trump (1946– ), "Trump: Art of the Deal"*

Sometimes a marvelous puzzle arises from the description of a
game. Is the game fair? What's the best strategy? An odd (actually, even) feature of the puzzles in this chapter is that each has
two versions, with entertaining contrasts between the two. There
are four pairs of games: The first involves numbers, the second hats,
the third cards, and the fourth gladiators.

We begin with a classic game which makes a great example in
a class on randomized algorithms (and indeed, was used that way
by Manuel Blum, now a professor at Carnegie Mellon University).

## Comparing Numbers, Version I

Paula (the perpetrator) takes two slips of paper and writes an integer
on each. There are no restrictions on the two numbers except that
they must be different. She then conceals one slip in each hand.

Victor (the victim) chooses one of Paula's hands, which Paula
then opens, allowing Victor to see the number on that slip. Victor
must now guess whether that number is the larger or the smaller of
Paula's two numbers; if he guesses right, he wins $1, otherwise he
loses $1.

Clearly, Victor can achieve equity in this game, for example, by
flipping a coin to decide whether to guess "larger" or "smaller."
The question is: Not knowing anything about Paula's psychology,
is there any way he can do better than break even?

## Comparing Numbers, Version II

Now let's make things much easier for Victor: Instead of being
chosen by Paula, the numbers are chosen independently at random

from the uniform distribution on [0,1] (two outputs from a standard random number generator will do fine).

To compensate Paula, we allow her to examine the two random numbers and *to decide which one Victor will see*. Again, Victor must decide whether the number he sees is the larger or smaller of the two, with $1 at stake. Can he do better than break even? What are his and Paula's best (i.e., "equilibrium") strategies?

# Red and Blue Hats, Version I

Each member of a team of $n$ players is to be fitted with a red or blue hat; each player will be able to see the colors of the hats of his teammates, but not the color of his own hat. No communication will be permitted. At a signal, each player will simultaneously guess the color of his own hat; all the players who guess wrong are subsequently executed.

Knowing that the game will be played, the team has a chance to collaborate on a strategy (that is, a set of rules—not necessarily the same for each player—telling each player which color to guess, based on what he sees). The object of their planning is to *guarantee* as many survivors as possible, assuming worst-case hat distribution.

In other words, we may assume the hat-distributing enemy knows the team's strategy and will do his best to foil it. How many players can be saved?

# Red and Blue Hats, Version II

Again, each of a team of $n$ players will be fitted with a red or blue hat; but this time the players are to be arranged in a line, so that each player can see only the colors of the hats in front of him. Again each player must guess the color of his own hat, and is executed if he is wrong; but this time, the guesses are made sequentially, from the back of the line toward the front. Thus, for example, the $i$th player in line sees the hat colors of players 1, 2, ..., $i-1$ and hears the guesses of players $n$, $n-1$, ..., $i+1$ (but he isn't told which of those guesses were correct—the executions take place later).

As before, the team has a chance to collaborate beforehand on a strategy, with the object of guaranteeing as many survivors as possible. How many players can be saved in the worst case?

# Betting on the Next Card, Version I

Paula shuffles a deck of cards thoroughly, then plays cards face up one at a time, from the top of the deck. At any time, Victor can interrupt Paula and bet $1 that the next card will be red. He bets once and only once; if he never interrupts, he's automatically betting on the last card.

What's Victor's best strategy? How much better than even can he do? (Assume there are 26 red and 26 black cards in the deck.)

# Betting on the Next Card, Version II

Again Paula shuffles a deck thoroughly and plays cards face up one at a time. Victor begins with a bankroll of $1, and can bet any fraction of his current worth, prior to each revelation, on the color of the next card. He gets even odds regardless of the current composition of the deck. Thus, for example, he can decline to bet until the last card, whose color he of course will know, then bet everything and be assured of going home with $2.

Is there any way Victor can *guarantee* to finish with more than $2? If so, what's the maximum amount he can assure himself of winning?

# Gladiators, Version I

Paula and Victor each manage a team of gladiators. Paula's gladiators have strengths $p_1, p_2, \ldots, p_m$ and Victor's, $v_1, v_2, \ldots, v_n$. Gladiators fight one-on-one to the death, and when a gladiator of strength $x$ meets a gladiator of strength $y$, the former wins with probability $x/(x+y)$, and the latter with probability $y/(x+y)$. Moreover, if the gladiator of strength $x$ wins, he gains in confidence and inherits his opponent's strength, so that his own strength improves to $x+y$; similarly, if the other gladiator wins, his strength improves from $y$ to $x+y$.

After each match, Paula puts forward a gladiator (from those on her team who are still alive), and Victor must choose one of his to face Paula's. The winning team is the one which remains with at least one live player.

What's Victor's best strategy? For example, if Paula begins with her best gladiator, should Victor respond from strength or weakness?

## Gladiators, Version II

Again Paula and Victor must face off in the Coliseum, but this time, confidence is not a factor, and when a gladiator wins, he keeps the same strength he had before.

As before, prior to each match, Paula chooses her entry first. What is Victor's best strategy? Whom should he play if Paula opens with her best man?

# Solutions and Comments

### Comparing Numbers, Version I

To the best of our knowledge, this problem originated with Tom Cover in 1986; see "Pick the Largest Number," in *Open Problems in Communication and Computation*, T. Cover and B. Gopinath, editors, Springer Verlag (1987), p. 152. Amazingly, there *is* a strategy which guarantees Victor a better than 50% chance to win.

Before playing, Victor selects a probability distribution on the integers that assigns positive probability to each integer. (For example, he plans to flip a coin until a "head" appears. If he sees an even number $2k$ of tails, he will select the integer $k$; if he sees $2k-1$ tails, he will select the integer $-k$.)

If Victor is smart, he will conceal this distribution from Paula, but as you will see, Victor gets his guarantee even if Paula finds out.

After Paula picks her numbers, Victor selects an integer from his probability distribution and adds $\frac{1}{2}$ to it; that becomes his "threshold" $t$. For example, using the distribution above, if he flips five tails before his first head, his random integer will be $-3$ and his threshold $t$ will be $-2\frac{1}{2}$.

When Paula offers her two hands, Victor flips a fair coin to decide which hand to choose, then looks at the number in that hand. If it exceeds $t$, he guesses that it is the larger of Paula's numbers; if it is smaller than $t$, he guesses that it is the smaller of Paula's numbers.

So, why does this work? Well, suppose that $t$ turns out to be larger than either of Paula's numbers; then Victor will guess "smaller" regardless of which number he gets, and thus will be right with probability exactly $\frac{1}{2}$. If $t$ undercuts both of Paula's numbers,

Victor will inevitably guess "larger" and will again be right with probability $\frac{1}{2}$.

But, *with positive probability*, Victor's threshold $t$ will fall *between* Paula's two numbers; and then Victor wins regardless of which hand he picks. This possibility, then, gives Victor the edge which enables him to beat 50%. ♡

Neither this nor any other strategy enables Victor to guarantee, for some fixed $\varepsilon > 0$, a probability of winning greater than 50% + $\varepsilon$. A smart Paula can choose randomly two consecutive multidigit integers, and thereby reduce Victor's edge to a smidgen.

## Comparing Numbers, Version II

It looks like the ability to choose which number Victor sees is paltry compensation to Paula for not getting to pick the numbers, but in fact *this* version of the game is strictly fair: Paula can prevent Victor from getting any advantage at all.

Her strategy is simple: Look at the two random real numbers, then feed Victor the one which is closer to $\frac{1}{2}$.

To see that this reduces Victor to a pure guess, suppose that the number $x$ revealed to him is between 0 and $\frac{1}{2}$. Then the unseen number is uniformly distributed in the set $[0, x] \cup [1-x, 1]$ and is, therefore, equally likely to be smaller or greater than $x$. If $x > \frac{1}{2}$, then the set is $[0, 1-x] \cup [x, 1]$ and the argument is the same.

Of course, Victor can guarantee probability $\frac{1}{2}$ against any strategy by ignoring his number and flipping a coin, so the game is completely fair. ♡

This amusing game was brought to my attention at a restaurant in Atlanta. Lots of smart people were present and were stymied, so if you failed to spot this nice strategy of Paula's, you're in good company.

## Red and Blue Hats, Version I

It is not immediately obvious that any players can be saved. Often the first strategy considered is "guessing the majority color"; e.g., if $n = 10$, each player guesses the color he sees on five or more of his nine teammates. But this results in ten executions if the colors are distributed five-and-five, and the most obvious modifications to this scheme also result in total carnage in the worst case.

However, it is easy to save $\lfloor n/2 \rfloor$ players by the following device. Have the players pair up (say, husband and wife); each husband chooses the color of his wife's hat, and each wife chooses the color she *doesn't* see on her husband's hat. Clearly, if a couple has the same color hat, the husband will survive; if not, the wife will survive.

To see that this is best possible, imagine that the colors are assigned uniformly at random (e.g., by fair coin-flips), instead of by an adversary. Regardless of strategy, the probability that any particular player survives is exactly $1/2$; therefore, the expected number of survivors is exactly $n/2$. It follows that the *minimum* number of survivors cannot exceed $\lfloor n/2 \rfloor$. ♡

## Red and Blue Hats, Version II

This version of Red and Blue Hats was passed on to me by Girija Narlikar of Bell Labs, who heard it at a party (the previous version was my own response to Girija's problem, but has no doubt been considered before). For the sequential version, it is easy to see that $\lfloor n/2 \rfloor$ can be saved; for example, players $n$, $n-2$, $n-4$, etc. can each guess the color of the player immediately ahead, so that players $n-1$, $n-3$, etc. can echo the most recent guess and save themselves.

It seems like some probabilistic argument such as the one provided for the simultaneous version should also work here, to show that $\lfloor n/2 \rfloor$ is the most that can be saved. Not so: All the players except the last can be saved!

The last player (poor fellow) merely calls "red" if he sees an odd number of red hats in front of him, and "blue" otherwise. Player $n-1$ will now know the color of his own hat; for example, if he hears player $n$ guess "red" and sees an *even* number of red hats ahead, he knows his own hat is red.

Similar reasoning applies to each player going up the line. Player $i$ sums the number of red hats he sees and red guesses he hears; if the number is odd, he guesses "red," if even, he guesses "blue," and he's right (unless someone screwed up).

Of course, the last player can never be saved, so $n-1$ is best possible. ♡

It is worth noting (thanks to Joe Buhler for mentioning this) that even if there are $k$ different hat colors instead of only two, only that last player in line need be sacrificed. He codes the colors as

$0, 1, 2, \ldots, k-1$ and adds the colors of all the hats he sees, modulo $k$. He then guesses the color corresponding to the sum, and now each other player can determine his hat color by subtracting from this first guess the sum of the colors he sees and subsequent guesses he has heard.

The last player's strategy (in the $k = 10$ case) might be exactly what is used by your bank to construct the check digit at the end of your account number.

## Betting on the Next Card, Version I

It looks as if Victor can gain a small advantage in this game by waiting for the first moment when the red cards in the remaining deck outnumber the black, then making his bet. Of course, this may never happen and if it doesn't, Victor will lose; does this compensate for the much greater likelihood of obtaining a small edge?

In fact, it's a fair game. Not only has Victor no way to earn an advantage, he has no way to lose one either: All strategies are equally ineffective.

This fact is a consequence of the martingale stopping time theorem, and can also be established without much difficulty by induction on the number of cards of each color in the deck. But there is another proof, which I will describe below, and which must surely be in "the book."[1]

Suppose Victor has elected a strategy S, and let us apply S to a slightly modified variation of "Betting on the Next Card, Version I." In the new variation, Victor interrupts Paula as before, but this time he is betting not on the *next* card in the deck, but instead on the *last* card of the deck.

Of course, in any given position, the last card has precisely the same probability of being red as the next card. Thus, the strategy S has the same expected value in the new game as it did before.

But, of course, the astute reader will already have observed that the new variation is a pretty uninteresting game; Victor wins if the last card is red, regardless of his strategy.

There is a discussion of "Betting on the Next Card, Version I" in the book *Elements of Information Theory* by T. Cover and J. Thomas, Wiley (1991), based on a result in T. Cover, "Universal

---

[1]As many readers will know, the late, great mathematician Paul Erdős often spoke of a book owned by God in which is written the best proof of each theorem. I imagine Erdős is reading the book now with great enjoyment, but the rest of us will have to wait.

Gambling Schemes and the Complexity Measures of Kolmogorov and Chaitin," Statistics Department Technical Report #12, Stanford University, October 1974.                                    ♡

The modified version of "Next Card Red" reminds me of a game which was described—for satiric purposes—in the *Harvard Lampoon*[2] many years ago. Called "The Great Game of Absolution and Redemption," it required that the players move via dice rolls around a Monopoly™-like board, until everyone has landed on the square marked "DEATH." So who wins?

At the beginning of the game, you are dealt a card face down from the Predestination Deck. At the conclusion, you turn your card face up, and if it says "damned," you lose.

### Betting on the Next Card, Version II

Finally, we have a really good game for Victor. But can he guarantee to do better than doubling his money, irrespective of how the cards are distributed?

It is useful first to consider which of Victor's strategies are optimal in the sense of "expectation." It is easy to see that as soon as the deck comes down to all cards of one color, Victor should bet everything at every turn for the rest of the game; we will dub any strategy which does this "reasonable." Clearly, every optimal strategy is reasonable.

Surprisingly, the converse is also true: No matter what Victor's strategy is, as long as he comes to his senses when the deck becomes monotone, his expectation is the same! To see this, consider first the following *pure* strategy: Victor imagines some fixed specific distribution of red and black in the deck, and bets *everything he has* on that distribution *at every turn*.

Of course, Victor will nearly always go broke with this strategy, but if he wins he can buy the earth—his take-home is then $2^{52} \times \$1$, around 50 quadrillion dollars. Since there are $\binom{52}{26}$ ways the colors can be distributed in the deck, Victor's mathematical expected return is $\$2^{52}/\binom{52}{26} = \$9.0813$.

Of course, this strategy is not realistic, but it is "reasonable" by our definition, and, most importantly, *every reasonable strategy is a combination of pure strategies of this type*. To see this, imagine that

---

[2]*Harvard Lampoon* Vol. CLVII, No. 1, March 30, 1967, pp. 14–15. The issue is dubbed "Games People Play Number" and the game in question appears to have been composed by D. C. Kenney and D. C. K. McClelland.

Victor had $\binom{52}{26}$ graduate students working for him, each playing a different one of the pure strategies.

We claim that every reasonable strategy of Victor's amounts to distributing his original $1 stake among these assistants, in some way. If at some point his collective assistants bet $x on red and $y on black, that amounts to Victor himself betting $x − $y on red (when $x > y$) and $y − $x on black (when $y > x$).

Each reasonable strategy yields a distribution, as follows. Say Victor wants to bet $.08 that the first card is red; this means that the assistants who are guessing "red" first get a total of $.54 while the others get only $.46. If, on winning, Victor plans next to bet $.04 on black, he allots $.04 more of the $.54 total to the "red-black" assistants than to the "red-red" assistants. Proceeding in this manner, eventually each individual assistant has his assigned stake.

Now, any convex combination of strategies with the same expectation shares that expectation, hence every reasonable strategy for Victor has the same expected return of $9.08 (yielding an expected profit of $8.08). In particular, all reasonable strategies are optimal.

But one of these strategies *guarantees* $9.08; namely, the one in which the $1 stake is divided equally among the assistants. Since we can never guarantee more than the expected value, this is the best possible guarantee. ♡

This strategy is actually quite easy to implement (assuming as we do that US currency is infinitely divisible). If there are $b$ black cards and $r$ red cards remaining in the deck, where $b \geq r$, Victor bets a fraction $(b-r)/(b+r)$ of his current worth on black; if $r > b$, he bets $(r-b)/(b+r)$ of his worth on red.

If the original $1 stake is *not* divisible, but is composed of 100 indivisible cents, things become more complicated and it turns out that Victor does about a dollar worse. A dynamic program (written by Ioana Dumitriu, now at UC Berkeley) shows that optimal play by Victor and Paula results in Victor ending with $8.08; the table below shows the size of Victor's bankroll at each stage of a well-played game. For example, if the game reaches a point when there are 12 black and 10 red cards remaining, Victor should have $1.08. By comparing the entries above and to the right, we see that he should bet either $.11 (in which case Paula will let him win) or $.12 (in which case he will lose) that the next card is black.

```
 0  ───────────────────────────────────────────────────  101 202 404 808
 1  ───────────────────────────────────────────────────  202 303 404 404
 2  ──────────────────────────────────────────────  190 253 303 303 202
 3  ───────────────────────────────────  96 134 178 222 253 253 202 101
 4  ──────────────────────────────────  132 167 200 222 222 190
 5  ──────────────────────────────  101 129 158 184 200 200 178
 6  ──────────────────────────  126 150 171 184 184 167 134
 7  ──────────────────────  123 144 161 171 171 158 132  96
 8  ──────────────────  101 120 138 153 161 161 150 129
 9  ──────────────  83 100 117 133 146 153 153 144 126 101
10  ──────────────  99 115 129 140 146 146 138 123
11  ────────────  98 112 125 135 140 140 133 120
12  ──────────  97 110 121 130 135 135 129 117 101
13  ────────  96 108 118 126 130 130 125 115 100
14  ──────  95 106 115 122 126 126 121 112  99  83
15  ────  94 104 113 119 122 122 118 110  98
16  ──  73  83  93 102 110 116 119 119 115 108  97
17  ──  83  92 101 108 113 116 116 113 106  96
18  ──  91  99 106 111 113 113 110 104  95
19  ──  98 104 109 111 111 108 102  94
20  74  82  90  97 103 107 109 109 106 101  93
21  82  89  96 101 105 107 107 104  99  92  83
22  89  95 100 103 105 105 103  98  91  83  73
23  94  99 102 103 103 101  97
24  98 101 102 102 100  96  90
25 100 101 101  99  95  89  82
26 100 100  98  94  89  82  74

    26 25 24 23 22 21 20 19 18 17 16 15 14 13 12 11 10  9  8  7  6  5  4  3  2  1  0
```

Note that Victor tends to bet slightly more conservatively in the "100 cents" game than in the continuous version. If instead he chooses always to bet the nearest number of cents to the fraction $(b-r)/(b+r)$ of his current worth, Paula will bankrupt him before half the deck is gone!

I heard this problem from Russ Lyons, of Indiana University, who heard it from Yuval Peres, who heard it from Sergiu Hart; Sergiu doesn't remember where he heard it, but suspects that Martin Gardner may have written about it decades ago.

### Gladiators, Version I

As in Version I of "Betting on the Next Card," all strategies for Victor are equally good.

To see this, imagine that strength is money. Paula begins with $P = p_1 + \cdots + p_m$ dollars and Victor with $V = v_1 + \cdots + v_n$ dollars. When a gladiator of strength $x$ beats a gladiator of strength $y$, the former's team gains \$$y$ while the latter's loses \$$y$; the total amount of money always remains the same. Eventually, either Paula will finish with \$$P$ + \$$V$ and Victor with zero, or the other way 'round.

The key observation is that every match is a fair game. If Victor puts up a gladiator of strength $x$ against one of strength $y$, then his expected financial gain is

$$\frac{x}{x+y} \cdot \$y \; + \; \frac{y}{x+y} \cdot (-\$x) \; = \; \$0.$$

Thus, the whole tournament is a fair game, and it follows that Victor's expected worth at the conclusion is the same as his starting stake, \$$P$. We then have

$$q(\$P + \$V) + (1 - q)(\$0) = \$P,$$

where $q$ is the probability that Victor wins. Thus, $q = P/(P+V)$, independent of anyone's strategy in the tournament.  ♡

Here's another, more combinatorial proof, devised by one of my favorite collaborators, Graham Brightwell of the London School of Economics. Using approximation by rationals and clearing of denominators, we may assume that all the strengths are integers. Each gladiator is assigned $x$ balls if his initial strength is $x$, and all the balls are put into a uniformly random vertical order. When two gladiators battle, the one whose topmost ball is highest wins (this happens with the required $x/(x+y)$ probability) and the loser's balls accrue to the winner.

The surviving gladiator's new set of balls is still uniformly randomly distributed in the original vertical order, just as if he had started with the full set; hence, the outcome of each match is independent of previous events, as required. But regardless of strategy, Victor will win if and only if the top ball in the whole order is one of his; this happens with probability $P/(P+V)$.

## Gladiators, Version II

Obviously, the change in rules makes strategy considerations in this game completely different from the previous one—or does it? No, again the strategy makes no difference!

For this game, we take away each gladiator's money (and balls), and turn him into a light bulb.

The mathematician's ideal light bulb has the following property: Its burnout time is completely memoryless. That means that knowing how long the bulb has been burning tells us absolutely nothing about how long it will continue to burn.

You may know that the unique probability distribution with this property is the exponential; if the expected (average) lifetime of the bulb is $x$, then the probability that it is still burning at time $t$ is $e^{-t/x}$. However, no formula is necessary for this puzzle. You only need to know that a memoryless probability distribution exists.

Given two bulbs of expected lifetimes $x$ and $y$, respectively, the probability that the first outlasts the second is $x/(x+y)$. To see this without calculus, consider a light fixture that uses one "type $x$" bulb and one "type $y$" bulb; every time a bulb burns out, we replace it with another of the same type. When a bulb does burn out, the probability that it is the $y$-bulb is a constant independent of the past. But that constant must be $x/(x+y)$, because over a long period of time, we will use $y$-bulbs and $x$-bulbs in proportion $x : y$.

Back in the Coliseum, we imagine that the matching of two gladiators corresponds to turning on their corresponding light bulbs until one (the loser) burns out, then turning off the winner until its next match; since the distribution is memoryless, the winner's strength in its next match is unchanged. Substituting light bulbs for the gladiators may be less than satisfactory for the spectators, but it's a valid model for the fighting.

During the tournament, Paula and Victor each have exactly one light bulb lit at any given time; the winner is the one whose total lighting time (of all the bulbs/gladiators on her/his team) is the larger. Since this has nothing to do with the order in which the bulbs are lit, the probability that Victor wins is independent of strategy. (Note: That probability is a more complex function of the gladiator strengths than in the previous game). ♡

The constant-strength game appears in K. S. Kaminsky, E. M. Luks, and P. I. Nelson, "Strategy, Nontransitive Dominance and the Exponential Distribution," *Austral. J. Statist.*, Vol. 26, No. 2 (1984), pp. 111–118. I have a theory that the other game came about in the following way: Someone enjoyed the problem and remembered the answer (all strategies equally good), but not the conditions. When he or she tried to reconstruct the rules of the game, it was natural to introduce the inherited-strength condition in order to make a martingale.

# ᛕlgorithms

*Achievement is largely the product of steadily raising one's levels of
aspiration and expectation.*
                              —*Jack Nicklaus (1940– ), "My Story"*

Many fascinating mathematical puzzles revolve around algorithms.
Usually, you (the victim) are presented with a "situation," together
with a collection of possible operations and a target state. You may
or may not be able to exercise choice in applying the operations.
You are asked: Can you reach the target state? Or perhaps: Can
you *avoid* reaching the target state? And sometimes: In how many
operations?

   Typically, the operation changes some aspect of the situation for
the better, while possibly losing ground elsewhere. How can you
determine whether the target is reachable?

   Here is a practice problem from the 1st All Russian Mathematical Olympiad, 1961.

## Signs in an ᛕrray

Suppose that you are given an $m \times n$ array of real numbers and
permitted, at any time, to reverse the signs of all the numbers in
any row or column. Prove that you can arrange matters so that all
the row sums and column sums are non-negative.

Solution: Flipping a row that has a negative sum will fix that
sum, but possibly ruin some column sums. How can you be sure
to make progress?

   This puzzle conforms to the first of the following classic paradigms. In an algorithmic puzzle, you are typically presented with
a "current situation," a "target state," and a set of "operations"
which you can use to modify a situation. You are asked to prove
one of these statements (but not necessarily told which):

   (1) There is a (finite) sequence of operations which reaches the
       target state;

| 2 | -3 | -1 | 0 | 3 | | *1* |
|---|---|---|---|---|---|---|
| -1 | 2 | -1 | -1 | 2 | | *1* |
| 1 | 3 | 2 | -4 | 2 | | *4* |
| 2 | -1 | 0 | 3 | 1 | | *5* |
| -3 | -2 | -2 | 4 | 4 | | *1* |
| *1* | *-1* | *-2* | *2* | *12* | | |

$\Longrightarrow$

| 2 | 3 | -1 | 0 | 3 | | *7* |
|---|---|---|---|---|---|---|
| -1 | -2 | -1 | -1 | 2 | | *-3* |
| 1 | -3 | 2 | -4 | 2 | | *-2* |
| 2 | 1 | 0 | 3 | 1 | | *7* |
| -3 | 2 | -2 | 4 | 4 | | *5* |
| *1* | *1* | *-2* | *2* | *12* | | |

(2) Any sequence of operations will eventually reach the target;

(3) Every sequence of operations reaches the target in the same number of steps;

(4) No sequence of operations can reach the target.

Your goal in algorithmic problems should be to find a parameter $P$—some kind of numerical rating of states—which somehow encapsulates progress toward the target.

To prove (1), you want to show that until the target is reached there is always an operation (or sequence of operations) available which improves $P$. To make sure that you don't get caught in Zeno's paradox (making smaller and smaller steps, and never reaching the target value), you may have to show that $P$ can always be improved by at least a certain amount, or that there are only finitely many possible situations.

To prove (2), you do the same except that now you show that *every* choice of operation improves $P$.

To prove (3), you show that every operation improves $P$ by the same amount.

To prove (4), you show that *no* operation improves $P$, yet attaining the target requires improvement.

Now let us return to the array problem. We see that the number of lines (rows and columns) with non-negative sum is the wrong parameter; this number could decrease even when a line with negative sum is flipped. Instead, let's try setting $P$ equal to the sum of all the entries in the array. Flipping a row with sum $-s$ increases $P$ by $2s$, since $P$ can be written as the sum of all the row sums (and similarly for columns). Since there are only finitely many reachable situations (actually, no more than $2^{m+n}$), and $P$ goes up every time you flip a negative-sum line, you must reach a time when all the line sums are non-negative.

This was a Type (1) problem, but as you see it could also have been phrased as a Type (2) problem, by specifying that only negative-sum lines may be flipped, then asking you to show that you *will* reach a point when all the line-sums are non-negative.

For the problems below, considerably more imagination may be required to find a parameter $P$ that works.

## The Infected Checkerboard

An infection spreads among the squares of an $n \times n$ checkerboard in the following manner: If a square has two or more infected neighbors, then it becomes infected itself. (Neighbors are orthogonal only, so each square has at most four neighbors.)

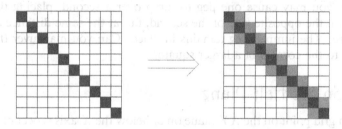

For example, suppose that we begin with all $n$ squares on the main diagonal infected. Then the infection will spread to neighboring diagonals and eventually to the whole board.

Prove that you can*not* infect the whole board if you begin with fewer than $n$ infected squares.

## Emptying a Bucket

You are presented with three large buckets, each containing an integral number of ounces of some nonevaporating fluid. At any time, you may double the contents of one bucket by pouring into it from a fuller one; in other words, you may pour from a bucket containing

$x$ ounces into one containing $y \leq x$ ounces until the latter contains $2y$ (and the former, $x-y$).

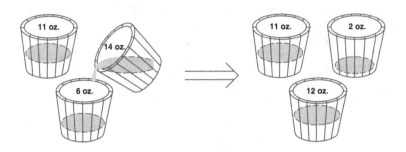

Prove that no matter what the initial contents, you can, eventually, empty one of the buckets.

## Pegs on the Corners

Four pegs begin on the plane at the corners of a square. At any time, you may cause one peg to jump over a second, placing the first on the opposite side of the second, but at the same distance as before. The jumped peg remains in place. Can you maneuver the pegs to the corners of a larger square?

## Pegs on the Half-Plane

Each grid point on the $XY$ plane on or below the $X$-axis is occupied by a peg. At any time, a peg can be made to jump over a neighbor peg (horizontally, vertically, or diagonally adjacent) and onto the

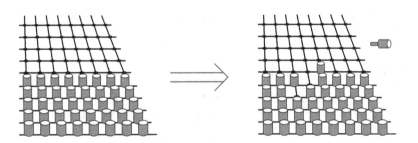

next grid point in line, provided that point was unoccupied. In this puzzle, however, the jumped peg is then removed.

Can you get a peg arbitrarily far above the $X$-axis?

## Pegs in a Square

Again we have pegs on the plane grid, this time in an $n \times n$ square. In this puzzle, pegs jump only horizontally or vertically, and the jumped peg is removed; the idea is to reduce your $n^2$ pegs to only 1.

Prove that if $n$ is a multiple of 3, it can't be done!

## Flipping the Polygon

The vertices of a polygon are labeled with numbers, the sum of which is positive. At any time, you may change the sign of a negative label, but then the new value is subtracted from both neighbors' values so as to maintain the same sum.

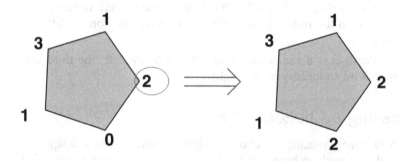

Prove that, inevitably, no matter which labels are flipped, the process will terminate after finitely many flips, with all values non-negative.

## Light Bulbs in a Circle

In a circle are light bulbs numbered 1 through $n$, all initially on. At time $t$, you examine bulb number $t$, and if it's on, you change the state of bulb $t+1$ (modulo $n$); i.e., you turn it off if it's on, and on if it's off. If bulb $t$ is off, you do nothing.

Prove that if you continue around and around the ring in this manner, eventually all the bulbs will again be on.

# Bugs on a Polyhedron

Associated with each face of a solid convex polyhedron is a bug which crawls along the perimeter of the face, at varying speed, but only in the clockwise direction. Prove that no schedule will permit all the bugs to circumnavigate their faces and return to their initial positions without incurring a collision.

# Bugs on a Line

Each positive integer on the number line is equipped with a green, yellow, or red light. A bug is dropped on "1" and obeys the following rules at all times: If it sees a green light, it turns the light yellow and moves one step to the right; if it sees a yellow light, it turns the light red and moves one step to the right; if it sees a red light, it turns the light green and moves one step to the *left*.

Eventually, the bug will fall off the line to the left, or run out to infinity on the right. A second bug is then dropped on "1," then a third.

Prove that if the second bug falls off to the left, the third will march off to infinity on the right.

# Breaking a Chocolate Bar

You have a rectangular chocolate bar marked into $m \times n$ squares, and you wish to break up the bar into its constituent squares. At each step, you may pick up one piece and break it along any of its marked vertical or horizontal lines.

Prove that every method finishes in the same number of steps.

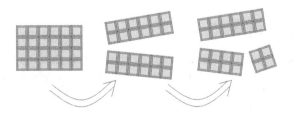

# Solutions and Comments

## The Infected Checkerboard

This lovely problem appeared in the Soviet magazine *KVANT* around 1986, then migrated to Hungary. When the initial squares are random, the process is called two-dimensional bootstrap percolation; a very nice mathematical analysis of the process has been done by Ander Holroyd (now of the University of British Columbia) and published in *Probability Theory and Related Fields*, Vol. 125, No. 2 (2003), pp. 195–224. The puzzle given here reached me through Joel Spencer of NYU, who claimed there was a "one-word proof"! As you will see, this is only a mild exaggeration.

Would-be solvers, misled by the diagonal example, often try to show that there must be an initially infected square in each row or column; but that is far from true. Note, for example, that the configuration of sick squares shown below spreads to the whole board.

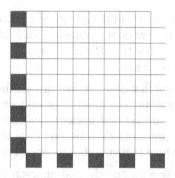

Indeed there are myriad ways to infect the whole board with $n$ sick squares, but apparently no way to do it with fewer. Some magic parameter $P$ is needed here, but what?

The parameter is the perimeter! When a square is infected, at least two of its boundary edges are absorbed into the interior of the infected area, and at most two added to the boundary of the infected area. Hence, the perimeter of the infected area cannot increase. Since the perimeter of the whole board is $4n$ (assuming unit-length edges), the initial infected area must have contained at least $n$ squares. ♡

An additional exercise for those interested: Prove that $n$ initial sick squares are necessary even when the top and bottom of the board are joined to form a cylinder. If the sides are joined as well, forming a torus, then $n-1$ initial sick squares are sufficient (and necessary). The perimeter no longer works, but another approach, found by Bruce Richter (University of Waterloo) and your author, does the trick.

## Emptying a Bucket

Yet another beauty from the former Soviet Union, this problem appeared in the 5th All Soviet Union Mathematical Olympiad, Riga, 1971. It showed up again, minus the hardware, on the Putnam Exam in 1993. The problem reached me via Christian Borgs of Microsoft Research. I will give two solutions: a combinatorial one of my own, and an elegant number-theoretic one found by Svante Janson of Uppsala University, Sweden (and independently by Garth Payne). I do not know which, if either, solution was the intended one.

In Svante's solution, $P$ is the content of a particular bucket and we show how $P$ can always be reduced until it is zero. In my solution, however, we show that $P$ can always be *increased* until one of the *other* buckets is empty.

To do the latter, we first note that we can assume there is exactly one bucket containing an odd number of ounces of fluid. This is true because if there are no odd buckets, we can scale down by a power of 2; if there are more than two odd buckets, one step with two of them will reduce their number to one or none.

Second, note that with an odd and an even bucket we can always do a reverse step, i.e., get half the contents of the even bucket into the odd one. This is because each state can be reached from at most one state, thus if you take enough steps, you must cycle back to your original state; the state *just before* you return is the result of your "reverse step."

Finally, we argue that as long as there is no empty bucket, the odd bucket's contents can always be increased. If there is a bucket whose contents are divisible by 4, we can empty half of it into the odd bucket; if not, one forward operation between the even buckets will create such a bucket.  ♡

Here is Svante's solution, in his own words:

"Label the buckets $A$, $B$, $C$ with, initially, $a$, $b$, and $c$ ounces of fluid, where $0 < a \le b \le c$. I will describe a sequence of moves leading to a state where the minimum of the three amounts is smaller than $a$. If this minimum is zero we are home, otherwise we relabel and repeat.

"Let $b = qa + r$, where $0 \le r < a$ and $q \ge 1$ is an integer. Write $q$ in binary form: $q = q_0 + 2q_1 + \cdots + 2^n q_n$ where each $q_i$ is 0 or 1 and $q_n = 1$.

"Do $n+1$ moves, numbered $0, \ldots, n$, as follows: In move $i$ we pour from $B$ into $A$ if $q_i = 1$ and from $C$ into $A$ if $q_i = 0$. Since we always pour into $A$, its content is doubled each time, so $A$ contains $2^i a$ before the $i$th move. Hence, the total amount poured from $B$ equals $qa$, so at the end there remains $b - qa = r < a$ in $B$. Finally, observe that the total amount poured from $C$ is at most

$$\sum_{i=0}^{n-1} 2^i a < 2^n a \le qa \le b \le c,$$

so there will always be enough fluid in $C$ (and in $B$) to do these moves." ♡

As far as I know, no one knows even approximately how many steps are required for this problem (in whatever is the worst starting state involving a total of $n$ ounces of fluid). My solution shows that order $n^2$ steps suffice, but Svante's does better, bounding the number by a constant times $n \log n$. The real answer might be still smaller.

## Pegs on the Corners

This cute puzzle was brought to my attention by Mikkel Thorup of AT&T Labs, who heard it from Assaf Naor (currently a post-doctoral researcher at Microsoft), who heard it from graduate students at Hebrew University, Jerusalem.

Note first that if the pegs begin on the points of a grid (i.e., points on the plane with integer coordinates), then they will remain on grid points.

In particular, if they sit initially at the corners of a unit grid square, then they certainly cannot later find themselves at the corners of a *smaller* square since no smaller square is available on the grid points. But why not a larger one?

Here's the key observation: The jump step is reversible! If you could get to a larger square, you could reverse the process and end up at a smaller square, which we now know is impossible.     ♡

## Pegs on the Half-Plane

This is a variation of a problem described in *Winning Ways*, Vol. 2; we believe the problem was invented originally by the second author, Conway. In his problem, diagonal jumps were not permitted; one can nonetheless get a peg to the line $y = 4$ without much difficulty, but an argument like the one below shows that no higher position can be reached.

With or without diagonal jumps, the difficulty is that as pegs rise higher, grid points beneath them are denuded. What is needed is a parameter $P$ which is rewarded by highly placed pegs, but compensatingly punished for holes left behind. A natural choice would be a sum over all pegs of some function of the peg's position. Since there are infinitely many pegs, we must be careful to ensure that the sum converges.

We could, for example, assign value $r^y$ to a peg on $(0, y)$, where $r$ is some real number greater than 1, so that the values of the pegs on the lower $Y$-axis sum to the finite number $\sum_{y=-\infty}^{0} r^y = r/(r-1)$. Values on adjacent columns will have to be reduced, though, to keep the sum over the whole plane finite; if we cut by a factor of $r$ for each step away from the $Y$-axis, we get a weight of $r^{y-|x|}$ for the peg at $(x, y)$, and a total weight of

$$\frac{r}{r-1} + \frac{1}{r-1} + \frac{1}{r-1} + \frac{1}{r(r-1)} + \frac{1}{r(r-1)} + \cdots = \frac{r^2 + r}{(r-1)^2} < \infty$$

for the initial position.

If a jump is executed, then at best (when the jump is diagonally upward and toward the $Y$-axis), the gain to $P$ is $vr^4$ and the loss $v + vr^2$, where $v$ is the previous value of the jumping peg. As long as $r$ is at most the square root of the "golden ratio" $\theta = (1 + \sqrt{5})/2 \approx 1.618$, which satisfies $\theta^2 = \theta + 1$, this gain can never be positive.

If we go ahead and assign $r = \sqrt{\theta}$, then the initial value of $P$ works out to about 39.0576; but the value of a peg at the point $(0, 16)$ is $\theta^8 \approx 46.9788$ *by itself*. Since we cannot increase $P$, it follows that we cannot get a peg to the point $(0, 16)$. But if we could get a peg to *any* point on or above the line $y = 16$, then we could get one to $(0, 16)$ by stopping when some peg reaches a point $(x, 16)$, then redoing the whole algorithm shifted left or right by $|x|$.

$\heartsuit$

We do not know the highest value of $y$ for which the point $(0, y)$ is reachable, allowing diagonal jumps. Perhaps an industrious reader can close this gap.

## Pegs in a Square

There is more than one way to solve this puzzle, which is *part* of a problem which appeared at the 1993 International Mathematical Olympiad. The proof given below was communicated to me by Benny Sudakov of Princeton University.

Color the points $(x, y)$ of the grid red if neither $x$ nor $y$ is a multiple of 3, otherwise white. This leaves a regular pattern of $2 \times 2$ squares (as in the figure).

If two pegs are (orthogonally) adjacent on the grid, both on red points or both on white, the peg remaining after the jump will be on white. If one is on red and the other on white, however, the peg remaining after the jump will be on red. It follows that if you start with any configuration having an even number of pegs on red squares, then this property will persist forever regardless of what jumps are made.

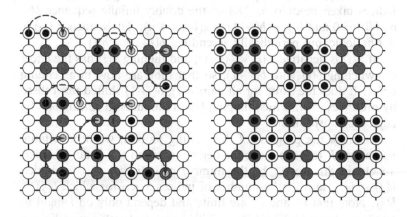

It is easy to see that a $3 \times 3$ square of pegs, no matter where it is placed on the plane grid, hits an even number of red points. Since an $n \times n$ square with $n$ a multiple of 3 is composed of such squares, it too will always hit an even number of red points. Suppose, however, that it were possible to reduce such a square to a single peg. Then we could shift the original square so that the surviving peg ended up on a red point, and this contradiction concludes the proof.

$\heartsuit$

It is routine, but not particularly easy or enlightening, to show that if $n$ is *not* a multiple of 3, you *can* reduce an $n \times n$ square to a single peg. At the Olympiad, contestants were asked to determine precisely for which $n$ the squares were reducible—pretty tough to do on the spot!

### Flipping the Polygon

This puzzle generalizes one that appeared at the International Mathematics Olympiad in 1986 (submitted by a composer from East Germany, I am told) and subsequently termed "the Pentagon Problem."

The problem has many solutions, and can even be generalized further, from $n$-gons to arbitrary connected graphs. However, the solution below stands out for its combination of elegance and strong conclusion. It was devised independently by at least two individuals, of whom one is Bernard Chazelle, Professor of Computer Science at Princeton University.

Let $x(0), \ldots, x(n-1)$ be the labels, summing to $s > 0$, with indices taken modulo $n$. Define the doubly infinite sequence $b(\cdot)$ by $b(0) = 0$ and $b(i) = b(i-1) + x(i \mod n)$. The sequence $b(\cdot)$ is not periodic, but periodically ascending: $b(i+n) = b(i) + s$.

If $x(i)$ is negative, $b(i) < b(i-1)$, and flipping $x(i)$ has the effect of switching $b(i)$ with $b(i-1)$, so that they are now in ascending order. It does the same for all pairs $b(j)$, $b(j-1)$ shifted from these by multiples of $n$. Thus, flipping labels amounts to sorting $b(\cdot)$ by adjacent transpositions!

To track the progress of this sorting process, we need a finite parameter $P$ that measures the degree to which $b(\cdot)$ is out of order. To obtain this, let $i^+$ be the number of indices $j > i$ for which $b(j) < b(i)$, and $i^-$ the number of indices $j < i$ for which $b(j) > b(i)$. Note that $i^+$ and $i^-$ are finite and depend only on $i \mod n$. Observe also that $\sum_{i=0}^{n-1} i^+ = \sum_{i=0}^{n-1} i^-$; we let this sum be our magic parameter $P$.

When $x(i+1)$ is flipped, $i^+$ decreases by 1, and every other $j^+$ is unchanged. Thus, $P$ goes down by *exactly 1*. When $P$ hits 0,

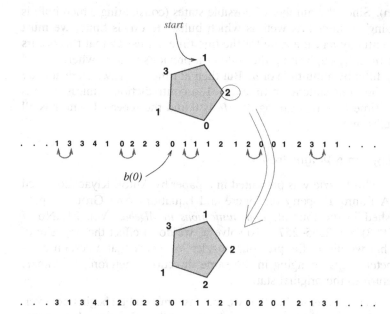

the sequence is fully sorted, so all labels are non-negative and the process terminates.

We have shown more than asked: The process terminates in exactly the same number ($P$) of steps regardless of choices, and moreover, the final configuration is independent of choices as well! The reason is that there is only one sorted version of $b(\cdot)$; entry $b(i)$ from the original sequence must wind up in position $i + i^+ - i^-$ when the sorting is complete. ♡

## Light Bulbs in a Circle

This puzzle is part of one that appeared at the International Mathematical Olympiad in 1993. With the value of $n$ unspecified, the best approach is to show (as we did in one of the "Emptying a Bucket" proofs) that the state space is itself a cycle.

We observe first that there is no danger of turning all the lights off; if a change is made at time $t$, bulb $t$ is still on. Moreover, if we look at the circle just *after* time $t$, we can deduce the state of the bulbs before $t$ (by changing the state of bulb $t+1$ if bulb $t$ is on). Since the number of possible states (considering which bulb is being examined as well as which bulbs are on) is finite, we must eventually repeat a state for the first time; let us say that this occurs at time $t_1$, duplicating the state at a previous time $t_0$ where $t_1$ and $t_0$ differ by a multiple of $n$. But then at time $t_1 - 1$, we were already in the same state as in time $t_0 - 1$, a contradiction—unless there is no time $t_0 - 1$, meaning that $t_0$ is 0 and the repeated state has all bulbs on. $\heartsuit$

## Bugs on a Polyhedron

This puzzle was presented in a paper by Anton Klyachko called "A Funny Property of Sphere and Equations over Groups," published in the journal, *Communications in Algebra*, Vol. 21, No. 7 (1993), pp. 2555–2575. To solve it, we do in effect the opposite of what we did in the previous puzzle: We show that a certain parameter keeps changing in the same direction, therefore, we can*not* return to the original state.

Let us observe first that we may assume no bug begins on a vertex (by advancing or retarding bugs slightly). We may also assume that the bugs move one at a time, crossing a vertex each time.

At any time, we may draw an imaginary arrow from the center of each face $F$, through $F$'s bug, to the center of the face on the other side of the bug. If we start at any face and follow these arrows, we must eventually hit some face a second time, completing a cycle of arrows on the polyhedron.

This cycle divides the surface of the polyhedron into two portions; let us define the "inside" of the cycle to be that portion surrounded clockwise by the cycle. Let $P$ be the number of vertices of the polyhedron inside the cycle.

Initially, $P$ could be anything from 0 to all ($n$, say) of the polyhedron's vertices; the extremes occur if there are two bugs on the same edge, causing a cycle of length 2. In the $P = 0$ case, the two bugs are facing each other, and doomed to collide.

When a bug on the cycle moves to its next edge, the arrow through it rotates to the right. The vertex through which it passed, previously on the inside of the cycle, is now outside; other vertices may also have passed from inside to outside the cycle, but there is no way for a vertex to move *inside*. To see this, note that the new arrow now points inside the cycle. The chain of arrows emanating from its head has no way to escape the cycle so must hit the tail

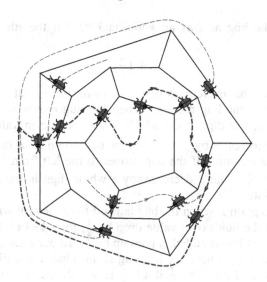

of some cycle arrow, creating a new cycle with smaller interior. In particular, $P$ has now dropped by at least 1.

Since we can never restore $P$ to its starting value, there is nothing to do but hope that the bugs are carrying collision insurance. ♡

## Bugs on a Line

We first need to convince ourselves that the bug will *either* fall of to the left, or go to infinity on the right; it cannot wander forever. To do so, it would have to visit some numbers infinitely often; let $n$ be the least of those numbers, but now observe that every third visit to $n$ will find it red and thus will incur a visit to $n-1$, contradicting the assumption that $n-1$ was visited only finitely often.

With that out of the way, it will be useful to think of a green light as the digit 0, red as 1, and yellow, perversely, as the "digit" $\frac{1}{2}$. The configuration of lights can then be thought of as a number between 0 and 1 written out in binary,

$$x = .x_1 x_2 x_3 \ldots,$$

where, numerically,

$$x = x_1 \cdot \left(\frac{1}{2}\right)^1 + x_2 \cdot \left(\frac{1}{2}\right)^2 + \cdots.$$

91

Think of the bug at $i$ as an additional "1" in the $i$th position, defining

$$y = x + \left(\frac{1}{2}\right)^i .$$

The point of this exercise is that $y$ is an *invariant*, that is, it does not change as the bug moves. When the bug moves to the right from point $i$, the digit upon which it sat goes up in value by $\frac{1}{2}$; therefore, $x$ increases by $\left(\frac{1}{2}\right)^{i+1}$, but the bug's own value diminishes by the same amount. If the bug moves to the left from $i$, it gains in value by $\left(\frac{1}{2}\right)^i$, but $x$ decreases by a whole digit in the $i$th place to compensate.

The exception is when the bug falls off to the left, in which case both $x$ and the bug's own value drop by $\frac{1}{2}$, for a loss of 1 overall. When the next bug is added, $y$ goes up by $\frac{1}{2}$. To put it another way, the value of $x$ goes up by $\frac{1}{2}$ if a bug is introduced and disappears to the right; and drops by $\frac{1}{2}$ if a bug is introduced and falls off to the left.

Of course, $x$ must always lie in the unit interval. If its initial value lies strictly between 0 and $\frac{1}{2}$, the bugs must alternate right, left, right, left; if between $\frac{1}{2}$ and 1, the alternation will be left, right, left, right.

The remaining cases can be checked by hand. If $x = 1$ initially (all points red) the first bug turns point 1 green and drops off to the left; the second wiggles off to infinity leaving all points red again, so the alternation is left, right, left, right. If $x = 0$ initially (all points green), the bugs will begin right, right (as the points change to all yellow, then all red), and then left, right, left, right as before.

The $x = \frac{1}{2}$ case is the most interesting because there are several ways to represent $\frac{1}{2}$ in our modified binary system: $x$ can be all $\frac{1}{2}$'s, or it can start with any finite number (including 0) of $\frac{1}{2}$'s, followed either by $0111\ldots$ or $1000\ldots$ . In the first case, the leadoff bug turns all the yellows to red as it zooms off to the right; thus, we get a right, left, right, left alternation. The second case is similar, the first bug wiggling off to the right, but again leaving all points red behind it. In the third case, the bug changes the yellows to red as it marches out, but when it reaches the red point, it reverses and heads left, turning reds to green on its way to dropping off the left end. Thereafter, we are in the $x = 0$ case, so the final pattern is left, right, right, left, right, left, right.

Checking back all the cases, we see that indeed, whenever the second bug went left, the third went right. ♡

This elegant analysis was done by Ander Holroyd (University of British Columbia) and Jim Propp (University of Wisconsin) at a meeting of the Institute for Elementary Studies in Banff, Alberta, 2003. The bug was proposed by Propp as a way to simulate deterministically a random walk on the non-negative integers in which steps are made, independently, to the left with probability 1/3 and to the right with probability 2/3. In such a walk, a given bug drops off to the left or proceeds to infinity on the right with equal probability; as we saw, the deterministic model gives a strict alternation instead, after the first couple of bugs. The argument can be generalized to other random walks.

## Breaking a Chocolate Bar

This ridiculously easy puzzle has been known to stump some *very* high-powered mathematicians for as much as a full day, until the light finally dawns amid groans and beatings of heads against walls. Risking accusations of sadism, I omit the solution.

# More Games

*Half this game is 90% mental.*
—*Danny Ozark, manager of the Phillies baseball team*

Analyzing a game often requires solving, in effect, two puzzles: finding a good strategy, and finding a good argument (or a good strategy for the second player) that shows that the former is best possible.

Sometimes, however, you can get away with much less. Consider the following innocent-looking puzzle.

## Chomp

Two players take turns biting off pieces of an $m \times n$ rectangular chocolate bar marked into unit squares. Each bite consists of selecting a square and biting off that square plus every remaining square above and/or to its right. Each player wishes to avoid getting stuck with the lower-left square, which is poisonous.

Prove that, if the bar contains more than one square, then the first player has a winning strategy.

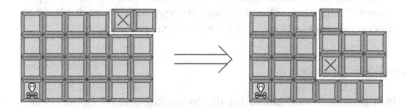

**Solution:** Either the first player (Alice) or the second (Bob) must have a winning strategy; suppose it is Bob. Then, in particular, Bob must have a winning answer to Alice's opening move when she merely nibbles off the upper right-hand square.

95

But whatever Bob's reply is could have been made by Alice as her own opening move, contradicting the assumption that Bob can always win. Hence, it must be Alice that has a winning strategy.♡

This kind of proof is known as a *strategy-stealing argument* and does not, unfortunately, tell you how Alice actually wins the game. More on Chomp, including its history and a more general version, will be found in the final chapter.

A variety of approaches will be useful for the rest of the game-puzzles.

## Deterministic Poker

Unhappy with the vagaries of chance, Alice and Bob elect to play a completely deterministic version of draw poker. A deck of cards is spread out face-up on the table. Alice draws five cards, then Bob draws five cards. Alice discards any number of her cards (the discarded cards will remain out of play) and replaces them with a like number of others; then Bob does the same. All actions are taken with the cards face-up in view of the opponent. The player with the better hand wins; since Alice goes first, Bob is declared to be the winner if the final hands are equally strong. Who wins with best play?

Deterministic poker is a full-information game. In games involving hidden information, or simultaneous moves, a randomized strategy may be called for. A set of such strategies (one for each player) is said to be in *equilibrium* if no player can gain by changing his strategy if the others keep theirs. For example, in "Rock, Paper, Scissors," the (unique) equilibrium strategy requires each player to choose among the three options with equal probability.

## Swedish Lottery

In a proposed mechanism for the Swedish National Lottery, each participant chooses a positive integer. The person who submits the lowest number not chosen by anyone else is the winner. (If no number is chosen by exactly one person, there is no winner.)

If just three people participate, but each employs an optimal, equilibrium, randomized strategy, what is the largest number that has positive probability of being submitted?

# Pancakes

Alice and Bob are hungry again, and now they are facing two stacks of pancakes, of height $m$ and $n$. Each player, in turn, must eat from the larger stack a (nonzero) multiple of the number of pancakes in the smaller stack. Of course, the bottom pancake of each stack is soggy, so the player who first finishes a stack is the loser.

For which pairs $(m, n)$ does Alice (who plays first) have a winning strategy?

How about if the game's objective is reversed, so that the first player to finish a stack is the winner?

# Determining a Difference

Alice and Bob relax after breakfast with a simple number game. Alternately, Alice chooses a digit and Bob substitutes it for one of the stars in the difference "$****$ $-$ $****$." Alice is trying to maximize the final difference, Bob to minimize it. What difference will be arrived at with best play?

# Three-Way Duel

Alice, Bob, and Carol arrange a three-way duel. Alice is a poor shot, hitting her target only $1/3$ of the time on average. Bob is better, hitting his target $2/3$ of the time. Carol is a sure shot.

They take turns shooting, first Alice, then Bob, then Carol, then back to Alice, and so on until only one is left. What is Alice's best course of action?

# Solutions and Comments

### Deterministic Poker

You need to know a little about the ranking of poker hands for this puzzle: namely, that the best type of hand is the straight flush (five cards in a row of the same suit), and that an Ace-high straight flush (also known as a "royal flush") beats a King-high straight flush and on down.

That means if Bob is allowed to draw a royal flush, Alice's goose is cooked. Thus, for Alice to have a chance, her initial hand must contain a card from each of the four possible royal flushes.

The best card of each suit, for that purpose, is the 10, since it stops all straight flushes which are 10-high or better. Indeed, a moment's thought will convince you that any hand of Alice's containing the four 10s will win. Bob cannot now hope to get a straight flush better than 9-high. To stop Alice getting a royal flush, he must draw at least one high card from each suit, thus only one card below a 10. Alice can now turn in four cards and make herself a 10-high straight flush in a suit other than the suit of Bob's low card, and Bob is helpless.                                                                   ♡

Alice has other winning hands as well. This odd game appeared in an early Martin Gardner column.

### Swedish Lottery

Suppose $k$ is the highest number any player is willing to play. If a player chooses $k$, he wins anytime the other two players agree, except if they agree on $k$. But if he chooses $k+1$, he wins anytime they agree, *period*. Hence, $k+1$ is a better play than $k$, and we cannot be in equilibrium. The contradiction shows that arbitrarily high submissions must be considered—sometimes one should choose 1,487,564.                                                         ♡

The actual equilibrium strategy calls for each player to submit the number $j$ with probability $(1-r)r^{j-1}$, where

$$r = -\frac{1}{3} - \frac{2}{\sqrt[3]{17 + 3\sqrt{33}}} + \frac{\sqrt[3]{17 + 3\sqrt{33}}}{3} \, ,$$

which is about 0.543689. The probabilities for choosing 1, 2, 3, and 4 are, respectively, about 0.456311, 0.248091, 0.134884, and 0.073335.

This rather nice lottery idea was brought to my attention by Olle Häggström of Chalmers University in Göteborg, Sweden. I do not know if it was ever implemented or even seriously considered for any official lottery, but don't you think it should have been?

### Pancakes

Suppose the stack sizes are currently $m$ and $n$, with $m > n$. If the ratio $r = m/n$ of stack sizes is strictly between 1 and 2, the next move is forced and the new ratio is $\frac{1}{1-r}$. These ratios are equal only for $r = \phi = (1+\sqrt{5})/2 \sim 1.618$, the golden mean; since $\phi$ is irrational, one of the two ratios $r$ and $\frac{1}{1-r}$ must exceed $\phi$ while the other is smaller than $\phi$.

The first player (Alice) wins exactly when the initial ratio of larger to smaller stack exceeds $\phi$. To see this, suppose $m > \phi n$, but $m$ is not a multiple of $n$. Write $m = an + b$, where $0 < b < n$. Then either $n/b < \phi$, in which case Alice eats $an$, or $n/b > \phi$, in which case she eats only $(a-1)n$. This leaves Bob with a ratio below $\phi$, and faced with a forced move which restores a ratio greater than $\phi$.

Eventually, Alice will reach a point where her ratio $m/n$ is an integer, at which point she can reduce to two equal stacks and stick Bob with a soggy pancake. But note that she can also, if desired, grab a whole pile for herself.

Of course, if Alice is instead faced with a ratio $m/n$ which is strictly between 1 and $\phi$, she is behind the eight-ball and it is Bob who can force the rest of the play.

We conclude that no matter which form of Pancakes is played, if the stacks are at heights $m > n$, Alice wins precisely when $m/n > \phi$. Only in the trivial case when the stacks are initially of equal height does it matter what the game's objective is. ♡

This puzzle, brought to my attention by Bill Gasarch of the University of Maryland, made an appearance on the 12th All Soviet Union Mathematical Olympiad, Tashkent, 1978.

## Determining a Difference

Write the difference as $x - y$, with $x = abcd$ and $y = efgh$. At any point in the game, let $x(0)$ be the result of substituting zeros for the remaining stars in $x$, and similarly for $x(9)$, $y(0)$, and $y(9)$. Alice guarantees at least 4000 by calling 5s and 4s until Bob puts a digit in position $a$, in which case Alice calls zeros for the rest of the game; or in position $e$, in which case she ends with all 9s. She must ensure that $x(9) \geq y(9)$ anytime she calls a "5" since Bob could place that 5 in position $e$, and similarly she must ensure that $x(0) \geq y(0)$ anytime she calls a "4" lest Bob put the 4 in position $a$. She can do this as follows.

Any time $x$ and $y$ are the same, Alice calls either "4" or "5." At any other point, let $u$ and $v$ be the symbols in $x$ and $y$, respectively, in the left-most of the position where $x$ and $y$ differ. If $u = *$ (in which case $x(9) > y(9)$), Alice calls "5"; if $v = *$ (in which case

$x(0) > y(0)$), she calls "4." It can never happen that $u = 4$ and $v = 5$, and if $u = 5$ and $v = 4$, both $x(9) > y(9)$ and $x(0) > y(0)$ hold so Alice can call either "4" or "5."

On the other hand, Bob ensures 4000 easily by immediately placing a 4 or lower in $a$, or a 5 or higher in $e$. He then leaves the other leading star alone while waiting for a non-0 (in the first case) or a non-9 (in the second). He thus achieves either $4000 - 0000$ or $9999 - 5999$, if not better. ♡

This puzzle goes back at least to the 6th All Soviet Union Mathematical Olympiad, Chelyabinsk, 1972.

## Three-Way Duel

I was reminded of this old chestnut by Dr. Richard Plotz of Providence, RI. It has appeared in many versions, one of which goes back at least to a 1938 puzzle book by Hubert Phillips called *Question Time*, published by J. M. Dent & Sons Ltd., London.

It's obvious that Alice should not be aiming at Bob; if she succeeds, she will subsequently be shot by Carol and that will be that.

Successfully shooting Carol will result in a two-way duel between Alice and Bob in which Bob has the better aim *and* the first move. Her chance of survival is clearly worse than $\frac{1}{3}$.

(In fact, if we let $p$ be her probability of survival when Bob begins and $q$ the (greater) probability of Alice's survival when she begins, we have $p = \frac{1}{3} \cdot q$ and $q = \frac{1}{3} + \frac{2}{3} \cdot p$, which gives $q = \frac{1}{7}$. Not good for Alice.)

If she misses, however, Bob will aim for Carol. If he succeeds, we are again faced with a duel between Alice and Bob, but this time Alice goes first, improving her odds to more than $\frac{1}{3}$ (in fact, to $\frac{3}{7}$).

If Bob fails, Carol will shoot him dead and Alice will get one chance to shoot Carol; her survival probability would in this case be $\frac{1}{3}$ exactly.

The point is that, regardless of whether Bob succeeds when aiming at Carol, Alice is better off missing than hitting Carol; and *much* better off missing than hitting Bob.

So, Alice's best strategy is to squander her first-up privilege by shooting into the air. ♡

# Handicaps

There are three side effects of acid. Enhanced long term memory,
decreased short term memory, and I forget the third.

*---Timothy Leary (1920--1996)*

In a popular Florida joke, Sam and Ted are two old codgers chatting
on Sam's front porch. "It's terrible," says Ted. "These days my
short-term memory is so poor, I can hardly remember each day
whether I've taken my daily pills or not."

"I know what you mean," replies Sam. "But my doctor has
found a solution—he added a special memory pill to my daily meds,
and it works wonders for me!"

"No kidding! What's the name of that pill? Maybe I can get
some, too!"

"Hmm, that's a good question. Let me think...umm...quick,
give me the name of a plant."

"A plant? You mean like a tree or bush?"

"No, something smaller, decorative..."

"A flower?"

"Yes, maybe a red one..."

"Carnation? Tulip?"

"No, it's got those prickly things..."

"Rose?"

"Yes! That's it!" Sam turns around and shouts through the
screen door. "Rose! What was the name of that memory pill?"

Algorithmic puzzles can impose bizarre handicaps, often having
to do with memory. It takes some imagination to deal with these
puzzles, and find a solution that can be applied by a less capable
being than yourself. In this company, our sample puzzle rates as
relatively realistic.

# Finding the Missing Number

All but one of the numbers from 1 to 100 are read to you, one every ten seconds, but in no particular order. You have a good mind, but only a normal memory, and no means of recording information during the process. How can you ensure that you can determine afterward which number was not called out?

Solution: Easy—you keep track of the sum of the numbers being called out, adding each one in turn to your accumulated total. The sum of *all* numbers from 1 to 100 is 100 times the average number ($50\frac{1}{2}$), namely 5050; that minus your final sum will be the missing number.

No need to keep the hundreds digit or thousands digit during the process, either; addition modulo 100 is good enough. At the end, you subtract the result from 50 or 150 to get an answer in the correct range. ♡

Dealing with streams of data, when handicapped by limited computing and memory resources, is a serious problem. Your first task is similar to the sample puzzle, but arose as a serious problem in the theory of computing.

# Identifying the Majority

A long list of names is read out, some names many times. Your object is to end up with a name that is guaranteed to be the name which was called a majority of the time, if there is such a name.

However, you have only one counter, plus the ability to keep just one name at a time in your mind. Can you do it?

The next puzzle was communicated to me by John H. Conway of Princeton University (inventor of the "Game of Life" on top of *many* other accomplishments). The problem is said to have immobilized a victim in his chair for six hours.

# The Conway Immobilizer

Three cards, an Ace, King, and Queen, lie face-up on a desk in some or all of three marked positions ("left," "middle," and "right"). If they are all in the same position, you see only the top card of the stack; if they are in two positions, you see only two cards and do not know which of the two is concealing the third card.

Your objective is to get the cards stacked on the left with Ace on top, then King, then Queen on bottom. You do this by moving one card at a time, always from the top of one stack to the top of another (possibly empty) stack.

The problem is, you have no short-term memory and must, therefore, devise an algorithm in which each move is based entirely on what you see, and not on what you last saw or did, or on how many moves have transpired. An observer will tell you when you've won. Can you devise an algorithm that will succeed in a bounded number of steps, regardless of the initial configuration?

Two of the remaining three puzzles involve light switches, very useful devices in puzzle composition. The last is a semiserious puzzle presaged by the opening joke.

## Spinning Switches

Four identical, unlabeled switches are wired in series to a light bulb. The switches are simple buttons whose state cannot be directly observed, but can be changed by pushing; they are mounted on the corners of a rotatable square. At any point, you may push, simultaneously, any subset of the buttons, but then an adversary spins the square. Show that there is a deterministic algorithm that will enable you to turn on the bulb in at most some fixed number of steps.

## The One-Bulb Room

Each of $n$ prisoners will be sent alone into a certain room, infinitely often, but in some arbitrary order determined by their jailer. The prisoners have a chance to confer in advance, but once the visits begin, their only means of communication will be via a light in the room which they can turn on or off. Help them design a protocol which will ensure that *some* prisoner will eventually be able to deduce that everyone has visited the room.

## The Two Sheriffs

Two sheriffs in neighboring towns are on the track of a killer, in a case involving eight suspects. By virtue of independent, reliable detective work, each has narrowed his list to only two. Now they are engaged in a telephone call; their object is to compare information, and if their pairs overlap in just one suspect, to identify the killer.

The difficulty is that their telephone line has been tapped by the local lynch mob, who know the original list of suspects but not which pairs the sheriffs have arrived at. If they are able to identify the killer with certainty as a result of the phone call, he will be lynched before he can be arrested.

Can the sheriffs, who have never met, conduct their conversation in such a way that they both end up knowing who the killer is (when possible), yet the lynch mob is still left in the dark?

## The Absent-Minded Pill Taker

An absent-minded professor of mathematics has to take a daily pill, but has problems with short-term memory and can never remember whether he has taken his pill for that day or not. To help himself, he has bought a transparent seven-day pill box with bins labeled SU, MO, TU, WE, TH, FR, SA. Fortunately, on account of his classes, the professor always knows what day of the week it is.

The problem is, he gets a new bottle of 30 or so pills whenever he runs out, and this could occur on any day of the week. He wants to empty the bottle completely into the pill box, but can't subsequently remember how many pills came in the bottle or on what day of the week he got the bottle.

The obvious approach of placing the pills in the box one at a time, starting with the current day, didn't work because when he later got to the point where the same number of pills remained in each bin, he couldn't determine whether he had taken that day's pill or not. The professor tried putting *all* the pills into the current day's bin, then moving them all to the right each time he took one. But he had trouble remembering to move them!

Can you supply the professor with an algorithm that will tell him, solely based on the day of the week and what he sees in the bins, whether he should take a pill and, if so, from which bin? The algorithm should tell him how to distribute the pills when they arrive, and should trust him to move any pills around later.

# Solutions and Comments

### Identifying the Majority

The idea is that whenever the counter is at 0 (where it begins), you put the name currently heard into your memory and increment the counter to 1. When the counter is greater than 0, increment it if the name you hear is the same as the one in memory; otherwise decrement it, but keep the same name in mind.

You could, of course, finish with a name in mind which occurred only once (e.g., if the list were "Alice, Bob, Alice, Bob, Alice, Bob, Charlie"). However, if a name occurs more than half the time, it's guaranteed to be the one in your memory at the end. The reason is that when this name is in memory, the counter is more often incremented than decremented. ♡

This algorithm is described in M. J. Fischer and S. L. Salzberg, "Finding a Majority Among $n$ Votes," *Journal of Algorithms* Vol. 3, No. 4 (December 1989), pp. 362–380.

## The Conway Immobilizer

It's tricky to design an algorithm that makes progress, avoids cycling, and doesn't do something stupid when it's about to win. The following will do the trick.

Move a card to the right (around the corner if necessary) to an empty slot, if there is an empty slot, unless you see K, –, A or K, A, –, in which case, place the Ace on the King. If all three cards are visible with the Queen on the left, place the King on the Queen; otherwise, move the card to the right of the Queen one space to the right (again, around the corner if necessary).

It's clear that no move produces a stack of three cards unless it is the winning configuration. Two-and-one configurations, even the ones that appear as K, –, A or K, A, –, will have all their cards exposed in at most three moves (unless the game is won). Thus, it suffices to check (see accompanying diagram) that the six possible configurations with all cards exposed lead to a win. ♡

Amazingly, the algorithm can be generalized to work with any (fixed, known) number of cards, still in three stacks. For, say, 52 cards numbered 1 through 52, the following rules (given in priority order) will eventually stack them in order on the left with 1 at the top:

(1) Seeing 2, 1, –, place 1 on 2;

(2) Seeing just two cards, move a card right (around the corner if necessary) to the open slot;

(3) Seeing $k, j, k-1$ with $j < k$, place $k-1$ on $k$;

(4) Seeing just one card, move a card to the left;

(5) Seeing three cards, move the card at the right of the largest numbered exposed card to the right.

Let us prove that this really works. Suppose card 52 is exposed in the center or right slot. Then using rules (2) and (5), it will eventually migrate to the left slot with all the rest of the cards stacked in the center. As these are moved to the right slot using rule (2), cards 51, 50, 49, ..., $k$ will be stacked on 52 via rule (3), for some $k < 52$, by the time the center slot is vacated. Of course, if $k = 1$, we are done, rule (1) having finished the job. Otherwise, card $k$ is then moved to the center by rule (2) and to the right by rule (5); card $k+1$ follows similarly, until 51 cards are stacked on the right with 51 down to $k$ on top.

Now 52 is moved to the center, the right stack inverted onto the left slot, 52 moved to the right, the left stack reinverted onto the center, and 52 moved back to the left. At this point, the center has 51 through $k$ on top, with 51 showing. Now cards 51 down to $k$ will each be moved right by rule (2) and then stacked on the left by rule (3), until, again, $k, k+1, \ldots, 52$ are stacked on the left.

The right slot is now empty, hence, card $k-1$ is now somewhere in the center. If it is not on the bottom, it will join the left stack and the above procedure will be repeated for $k' < k$. If it happens to be on the bottom, it will not be transferred to the left (unless it is card 1), because when it moves to the right via rule (5), the middle will be open, forcing us to use rule (2) instead of rule (3). However, the next time through, the center stack will be inverted, with $k-1$ on top. Hence, at least every other time that 52 is newly moved to an empty left slot, the value of $k$ drops.

Now we are done if we can show that the condition presupposed above--that 52 is exposed in the center or right slot--must eventually occur. Suppose first that 52 is exposed on the left (with other cards underneath). Then it may accumulate 51, 50, ..., $k$ on top of it via rule (3), ending with some cards in the center slot and possibly some on the right. Rules (2) and (5) will then clear the center slot. Card $k$ will move to the center and then to the right, then $k+1$ similarly, etc. (as above) until 52 is re-exposed--but this time (after 51 is moved to the right), the center slot will be empty. Card 52 will then be moved into the center, creating the desired condition.

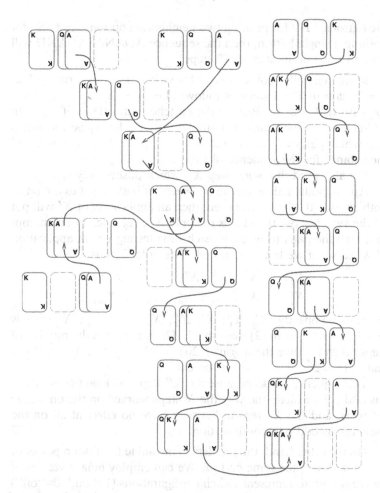

## Spinning Switches

This puzzle reached me via Sasha Barg of the University of Maryland, but seems to be known in many places. As with many puzzles, looking at a simpler version first helps. Consider the two-switch version: Pushing both buttons will ascertain if they were both in the same state, since then the bulb will light (if it wasn't already lit). Otherwise, we proceed to push one button, after which they *will* be in the same state, and at worst one more operation of pushing both buttons will turn on the bulb.

Back to the four-switch case. If we let "A" stand for the action of pushing all four buttons, "D" for pushing two diagonally opposite buttons, "N" for pushing two neighboring buttons, and "S" for pushing a single button, then the sequence ADANASADAND will turn on the bulb in at most 12 steps.

More generally, you can do it for switches at the corners of an $n = 2^k$-gon in $2^n - 1$ steps as follows: Let $X = X_1, \ldots, X_m$ be the steps for $n/2 = 2^{k-1}$. Pair up the switches antipodally; if $X_i$ is an $n/2$-gon step which pushes buttons $i_1, \ldots, i_j$, let $X_i'$ be the $n$-gon step which pushes $i_1, \ldots, i_j$ and $i_1 + n/2, i_2 + n/2, \ldots, i_j + n/2$. $X'$ then stands for the sequence of steps $X_1', \ldots, X_m'$.

We also need the $n$-gon step $X_i''$ which pushes only $i_1, \ldots, i_j$.

An antipodal pair is said to be "even" if both switches are on or both are off. If all pairs are even, then an application of $X'$ will put all the switches on; the idea is to apply $X_1''$, $X_2''$, etc., in an attempt to get all the pairs to be even, each time testing via an application of $X'$ to see if we have succeeded. The order is thus

$$X_1', \ldots, X_m'; \; X_1''; \; X_1', \ldots, X_m'; \; X_2'';$$
$$X_1', \ldots, X_m'; \; \ldots \; ; \; X_m''; \; X_1', \ldots, X_m',$$

or, more compactly, $X'; \; X_1''; \; X'; \; X_2''; \; X'; \; \ldots \; ; \; X_m''; \; X'$. This is $(m+1)m + m = m(m+2)$ steps in all. Then, if $f(n)$ is the number of steps taken to solve the $n$-gon, $f(2n) = (2^n - 1)(2^n + 1) = 2^{2n} - 1$ and $f(1) = 2^1 - 1 = 1$, so this checks.

The sequence works because the $X''$ steps work on the even versus odd pairs exactly the way the $X$ steps worked on the on versus off pairs, and the $X$ steps in between have no effect at all on the even versus odd pair configuration. ♡

On the other hand, the problem is insoluble for $n$ not a power of 2, say $n = m2^k$ for some odd $m$. We can employ binary vectors of length $n$ both to represent switch configurations (1="on," 0="off") and moves (1="push," 0="leave alone"). If $v$ is such a vector, we let $v^i$ represent the result of rotating $v$ $i$ steps to the right. Applying move $w$ to configuration $u$ would result in configuration $u + w$ if there were no spin, but since there is, we actually get $u + w^i$ for some unknown $i$.

Call an $n$-vector $v$ "rough" if the size of the set of its rotations $v = v^0, v^1, \ldots, v^{n-1}$ is not a power of two. Suppose that (as in the beginning) any rough configuration is possible in some rotation. Then we claim that after any fixed move $w$, any rough configuration is *still* possible in some rotation. Thus, you can never eliminate any of the rough configurations and in particular never guarantee to achieve $11 \ldots 1$.

For $n$ odd, i.e., $n = m$, all vectors except $00 \ldots 0$ and $11 \ldots 1$ are rough. If $w$ is any vector and $v$ is any rough vector, either $v - w$ (same as $v + w$) or $v - w^1$ will be rough, so if we had a rotation of one of those before we applied $w$, we might have a rotation of $v$ now.

If $n = m2^k$ for $k > 1$, we can break up the $n$-cycle into $m$ segments of length $2^k$, and $u$ will be rough as long as it is not the same on every segment. Thus, if it is rough, there is some $1 \leq j \leq 2^k$ such that coordinates $i2^k + j$, for $1 \leq i \leq m$, are not all equal. Now we apply the same argument as above, just looking at these $m$ coordinates.

## The One-Bulb Room

I heard this puzzle from Adam Chalcraft, who has the distinction of having represented Great Britain internationally in unicycle hockey. The puzzle has also appeared on www.ibm.com and was reprinted in *Emissary*, the newsletter of the Mathematical Sciences Research Institute in Berkeley, California. A version even appeared on the justly famous public radio program, *Car Talk*, in 2003.

However, the reader is warned that the puzzle has sometimes been confused with a much tougher problem, which he or she will encounter in the next chapter.

It will, of course, be necessary to assume that no one fools with the room's light between visits by prisoners; but prisoners do not need to know the initial state of the light. The idea is that one prisoner (say, Alice) repeatedly tries to turn the light on, and each of the others turns it off *twice*.

More precisely, Alice always turns the light on if she finds it off; otherwise, she leaves it on. The rest of the prisoners turn it off the first two times they find it on, but otherwise leave the light alone.

Alice keeps track of how many times she finds the room dark after her initial visit; after $2n - 3$ dark revisits, she can conclude that everyone has visited. Why? Every dark revisit signals that one of the other $n - 1$ prisoners has visited. If one of them, say Bob, hasn't been in the room, then the light cannot have been turned off more than $2(n-2) = 2n-4$ times. On the other hand, Alice *must* eventually achieve her $2n - 3$ dark revisits because eventually the light will have been turned off $2(n-1) = 2n-2$ times and only one of these (caused by a prisoner darkening an initially light room before Alice's first visit) can fail to cause a dark revisit by Alice.

If there are just two prisoners, it's clear that each can learn of the other's visit, since Alice can wait for her first dark revisit while Bob waits for his first *light* revisit. However, it is possible to show that for $n > 2$, there is no way to guarantee that more than one prisoner will get the word that everyone has been there. A sketchy proof is provided below; readers are encouraged to skip it unless they are particularly interested in how negative results can be obtained in a communications puzzle of this sort. I include the proof because, as far as I know, it appears nowhere else.

Basically, we argue that the adversary (who, we may assume, schedules the visits, knowing the prisoners' strategy) can render useless actions other than the ones used in the above protocol.

Let us focus on one prisoner, say Alice. Her strategy can be assumed to be deterministic and based solely on the sequence of light-states she has so far observed.

Suppose that Alice's strategy calls for her (in some circumstance) to change the state (of the light) after finding it in the state in which she last left it. Then the adversary could have brought her back immediately to the room, "wasting" her previous visit; in effect, this piece of Alice's strategy can only give the adversary an extra option. We may assume, therefore, that Alice never changes the state when she finds it where she last left it.

Next, suppose Alice is required at some point to leave the state as she found it. Then we claim we can assume she will never act again! Why? Because if the adversary doesn't want her ever to act again, he can insure that she never sees a state different from the state she now finds. He can do this because if Alice *did* become permanently inactive, at least one of the states (on or off) will recur infinitely often; suppose it's the "off" state. Then he can schedule Alice so that she sees "off" now and at every subsequent visit, hence, by the previous argument, she will never act again. So, once more, the adversary always has the option of silencing Alice so we may assume that it is his only option.

Obviously, Alice can't then begin with the instruction to leave the state as she finds it, since in that case she is forever inactive and no one will ever know that she has visited the room.[3] Say she is supposed to turn the light on if it's off, otherwise leave it on. Then she won't be doing anything until she again finds the light off, at which point she may only turn it on again or go inactive forever. Thus, she is limited to turning the light on some number of times

---

[3]Unless, of course, she wears a strong perfume ...

"$j$" (which may as well be constant, else the adversary has more options). We call this strategy $+j$, where $j$ is a positive integer or infinity. Similar arguments apply if she is instructed to turn the light off on first visit, leading to strategy $-j$.

The only remaining possibility is that she is instructed to change the state of the light at first visit, in which case she must proceed as above depending on whether she turned the first light on or off. This again only gives the adversary an additional option.

We are reduced to each prisoner having a strategy $+j$ or $-j$ for various $j$. If they all turn lights only off (or only on), no one will learn anything; thus, we may assume Alice's strategy is $+j$ and Bob's is $-k$. If Charlie turns lights on, Alice will never be able to tell the difference between Bob and Charlie both having finished, and Bob and Charlie each having one task left. If Charlie turns lights off, it's Bob who will be "left in the dark."

Putting all this together, we have that for a prisoner to be able to determine that everyone has visited, she must turn the light on, while everyone else turns it off (or vice-versa). In fact, if her strategy is $+j_1$ and the others are $-j_2, \ldots, -j_n$, then it's easy to check that having each $j_i$ finite, but at least 2, and $j_1$ greater than the sum of the other $j_i$'s minus the least of them, is necessary and sufficient.

It follows that if $n > 2$, then at most one prisoner can be guaranteed the privilege of knowing that all others have visited the room. Whew!

## The Two Sheriffs

If the two sheriffs (let us call them Lew and Ralph) share some secret information, they can use that secret to "encrypt" their conversation and achieve the objective. But not having met before, they will, in effect, need to manufacture their own secret.

Let us assume throughout that the pairs of suspects to which Lew and Ralph have narrowed their search are not identical, so that the killer is potentially identifiable. Notice that if Lew (say) merely names his pair, then Ralph will know the killer. But then the lynch mob will know everything Lew knows, so any effort by Ralph to communicate the name of the killer to Lew without also giving it to the mob must come to naught.

Apparently, Lew and Ralph will have to converge on the killer's name in a more subtle manner. Let us make a table of all $8 \cdot 7/2 = 28$ possible suspect pairs, in such a way that each column of the table constitutes a *partition* of the eight suspects into four pairs. Here is one way to do it:

$$\{1,2\} \quad \{1,3\} \quad \{1,4\} \quad \{1,5\} \quad \{1,6\} \quad \{1,7\} \quad \{1,8\}$$
$$\{3,4\} \quad \{2,4\} \quad \{2,3\} \quad \{2,6\} \quad \{2,5\} \quad \{2,8\} \quad \{2,7\}$$
$$\{5,6\} \quad \{5,7\} \quad \{5,8\} \quad \{3,7\} \quad \{3,8\} \quad \{3,5\} \quad \{3,6\}$$
$$\{7,8\} \quad \{6,8\} \quad \{6,7\} \quad \{4,8\} \quad \{4,7\} \quad \{4,6\} \quad \{4,5\}$$

Over the phone, Lew and Ralph are free to discuss the whole issue of keeping information from the lynch mob. In particular there is nothing to prevent them from agreeing on a numbering of the eight suspects, and arriving at a table such as this one.

Now, Lew tells Ralph which column his pair appears in. For example, if Lew's pair is $\{1,2\}$, he says "My pair is in the first column."

If Ralph's pair is in the same column, he immediately knows that he and Lew have the same pair. He can go ahead and say so, after which the sheriffs may as well hang up and go back to work.

Otherwise, Ralph knows that Lew's pair is one of two in that column; continuing the example, if Ralph's pair is $\{2,3\}$, he knows that Lew's pair must have been either $\{1,2\}$ or $\{3,4\}$. He then divides the column into two equal pieces, in such a way that both of these pairs are in the same piece, and announces the division to Lew.

In the example case, he might say to Lew "Either my pair is among $\{1,2,3,4\}$ or it is among $\{5,6,7,8\}$." (If, instead, Ralph's pair had been $\{2,5\}$, he would have said "Either my pair is among $\{1,2,4,5\}$ or it is among $\{3,4,7,8\}$.")

Lew will know, of course, which of the pieces is the one in which Ralph's pair is found, because it can only be the same piece in which his *own* pair is found. Lew and Ralph now share a secret!

Lew can now tell Ralph whether Lew's pair is the first or second inside the relevant piece. If, for example, the two pairs are $\{1,2\}$ and $\{1,3\}$ as above, Lew can say "My pair is the first in the piece," or, equivalently, "My pair is either $\{1,2\}$ or $\{5,6\}$."

Ralph then knows Lew's pair and therefore the killer's identity. He can communicate this knowledge simply by telling Lew whether the lower-numbered or higher-numbered suspect in Lew's pair is the killer. Here, he would say "The killer is the higher guy in your pair" or, equivalently, "The killer is either 2 or 6."

The lynch mob cannot know which "piece" Lew and Ralph are talking about. In the given example, the entire conversation heard by the mob would have been exactly the same had Lew's pair been $\{5,6\}$ and Ralph's $\{6,7\}$ or $\{6,8\}$, in which case the killer would have been 6 instead of 2. ♡

The Two Sheriffs puzzle appeared in D. Beaver, S. Haber, and P. Winkler, "On the Isolation of a Common Secret," from *The Mathematics of Paul Erdős* Vol. II, R. L. Graham and J. Nešetřil, editors, Springer-Verlag, Berlin, 1996. The puzzle was devised to give an example of a discovery your author made about twenty-five years ago: namely, that common knowledge can be molded into a common secret over an open channel. The initial application of this idea was to the game of bridge, where partners are not permitted to have prior private understandings about the meaning to each other of a bid or play. Since 1924 when contract bridge was invented, this rule was believed, incorrectly, to prohibit any secret communication within a partnership. This misperception had a chilling effect on the development of sophisticated methods for bidding and defense, because many players felt that such methods would give away too much information to the opponents; for example, a scientifically conducted slam auction would tell the opponents what to lead.

However, the cards in your own hand (which you know not to be held by your partner) give you and your partner common knowledge which can be used to communicate in secret. For details and references, see P. Winkler, "The Advent of Cryptology in the Game of Bridge," *Cryptologia* Vol. 7 #4 (October 1983), pp. 327–332.

## The Absent-Minded Pill Taker

The professor's pill box (pictured below) is composed of seven transparent bins, labeled SU, MO, TU, WE, TH, FR, SA. As an example of his problem, suppose that the professor begins with 30 fresh pills on a Friday morning. He wishes to distribute these among the bins in such a way that, by looking at the pill box starting today, he can tell whether he has taken his daily pill and, if not, can take one from an appropriate bin.

The obvious way to do this is to put five pills in FR and SA and four each in SU, MO, TU, WE, and TH. His algorithm is as follows:

When the pill box constitutes a contiguous (mod 7) string of bins with $k$ pills and the rest with only $k-1$, he knows that the leftmost "heavy" bin (with $k$ pills) is the place to go for the next pill. If it's Wednesday and that bin is marked WE he takes a pill from there, but if it's marked TH he knows that he's already taken his Wednesday pill.

The difficulty is that every seven days the bins become level, that is, each containing the same number of pills. Which bin is the chosen one then? Well, in this instance, the professor can see that this will happen on Sunday, so he decrees that when the bins are level, SU is the designated dispenser. So if he sees level bins and it's a Sunday, he takes from the SU bin; if it's Saturday, he's already taken his pill.

Things proceed nicely until 30 days later when he gets a new batch of 30 pills. Now it's a Sunday so if he distributes the pills with five each in SU and MO and the rest of the days with four, he can see that he will be level on Tuesday mornings, not on Sunday mornings. This would be a disaster; he'll never be able to remember that now it's TU that's the designated "level bin" and no longer SU.

Surely there's a way to solve this problem which doesn't involve leaving pills behind in the bottle (or throwing them out). But how? He needs some reliable way to mark the bin which is next to dispense. He could put a noticeably large number of pills in one bin and then move this "hoard" each day, but then he can't be sure that he's remembered to move the hoard. Somehow the professor has to devise an algorithm that allows him to just take his daily pill, and not move pills around.

Naturally, the professor begins to wonder if this is a solvable problem. Can he not construct an impossibility proof? If every day all he does is take a pill from the bin marked with that day, then, proceeding backward from the last pill, it's clear that every day the pill box is either level or has a contiguous set of bins with $k$ pills and the rest with $k-1$. Thus, he is back to his original problem of having to change and remember which bin is the designated "level bin."

But wait: Is there any reason why, say, Wednesday's pill has to be taken from the WE bin? Not really. Of course, the algorithm has to be kept simple otherwise even the professor's *long*-term memory might be taxed. But as long as there's a reasonable rule for which bin to take a pill from (as well as for determining whether the day's pill has already been taken), this extra flexibility is available.

As it turns out, there is an algorithm which meets all the professor's criteria and admits only one tiny exception to always taking from the bin marked by the current day of the week. The professor reasons as follows.

(1) The pill distribution in the pill box needs to be kept *fairly* level, since when the time for purchasing a new bottle nears, there won't be many pills left with which to play.

(2) The *completely* level distribution must be avoided, otherwise the problem of designating a "level bin" reappears.

(3) On account of (2), it cannot be right always to take from a bin with the maximum number of pills.

Considering these points, the professor arrives at the idea of maintaining at most three bin sizes at any time, and taking from a middle-sized bin when possible. To keep things as simple as possible, there will be only one "hoard"--a single bin containing the largest number of pills. On any given day, we'll let $k$ stand for the number of pills in the hoard. All other bins will contain $k-1$ or $k-2$ pills, and those that contain $k-2$ will be in contiguous bins which proceed to the right from the hoard. The figure below shows various legal configurations.

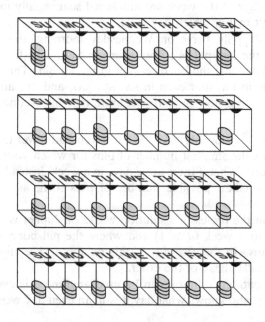

The first bin to the right of the hoard containing $k-1$ pills will be the one designated to dispense; if there is no bin with $k-1$ pills, then it's the hoard's turn to dispense. In (almost) every case, the dispensing bin is correctly marked with the current day of the week.

Thus, for example, the pill boxes in the figure are prepared to dispense on Tuesday, Saturday, Monday, and Thursday, respectively.

The exception comes when the professor is down to his last pill. On the previous day, he found two pills left, together in the bin marked by that day of the week; he took one (using the rule that when there is no bin of size one less than the hoard, he takes from the hoard). Now the last pill is in the bin marked for *yesterday* and it is that pill which he takes today.

It's easy to see that if the pills are properly distributed, then the configuration remains legal down to the last pill. But is it always possible to set up this scheme properly, when the pills arrive? Indeed, given any number of pills, and any morning of the week, there is a unique correct configuration; and it is that configuration which the professor constructs when the pills arrive. The professor simply calculates the day of the week on which the last pill will be taken (namely, yesterday's day of the week plus the number of pills, modulo 7—assuming today's pill hasn't been taken yet). Of course, the days of the week are numbered sequentially modulo 7, but it doesn't matter which day is "1."

If, say, 32 pills arrive on Wednesday morning, the professor knows that the last pill is to be taken on Saturday (from the FR bin!). It follows that the hoard is kept in the FR bin. The professor puts six pills in FR, four each in SA, SU, MO, and TU, and five in WE and TH. He is now properly set up to take his Wednesday pill.

♡

One might reasonably ask: "What if you had fewer than seven bins? What's the smallest number of bins for which your problem has a solution? What if there are $d$ days in a week, instead of seven; what's the smallest possible number of bins then, as a function of $d$?"

Note that the professor's solution works on Jupiter where there are $d$ days to a week ($d > 1$) and where the pill boxes have, of course, $d$ bins. In the case $d = 2$, this reduces to keeping one bin one or two pills ahead of the other.

But the two-bin solution can be used anytime $d$ is even, since the pill taker who knows what day it is in an even-day week knows the parity of his day. So two bins are sufficient, and of course necessary, when $d$ is even.

When $d$ is odd, however, two bins will not work. There must be two consecutive days of the week which are slated to come down to the same one-pill configuration, so when the pill taker sees this configuration on the first of those two days, he cannot know whether he has taken that day's pill or not.

The reader who has gotten this far will not find it difficult to convince him- or herself that three bins are sufficient when $d$ is odd. It's a little tricky to come up with a simple algorithm for three bins with a seven-day week; the following scheme makes mnemonic use of binary notation.

Let us agree on a numbering of the days of the week beginning with Sunday=1 and ending with Saturday=7, with the numbers taken modulo 7. The scheme involves seven configuration "types," named 1 through 7, where the shape of each type corresponds to the binary representation of its name. The bins themselves are linearly placed ("left," "center," and "right") and not considered cyclically.

Thus, for example, type $1 = 001_2$ demands that the rightmost bin act as a hoard with noticeably more pills than either of the other two bins. Type $3 = 011_2$ demands that the leftmost bin have substantially *fewer* pills than the other two bins; and Type $7 = 111_2$ demands that the bin occupancies be kept nearly level.

More exactly, types 1, 2, and 4 have a hoard (on the right, center, and left, respectively) with two or three pills more than the other two bins; the other two, if different, are ordered with the larger bin toward the right.

Types 3, 5, and 6 have a unique smallest bin on the left, center, or right, respectively; the other two are larger by 2 if they are the same. If they are not the same they differ in size by at most 1 with the larger toward the right, and have two or three more pills than the small bin.

Type 7 demands that the occupancies all be within one pill of one another with the smaller bins on the right. (See the table on the following page.)

The strategy is now as follows: If $P$ pills arrive on day $D$, they are distributed in accordance with type $D + P$ mod 7. The pill taker then takes pills so as to maintain the type.

In particular, each day, the pill taker observes the type $T$ and operates as follows: If he sees $P > 3$ pills on day $D$ and $D + P \neq T$ mod 7, he has already taken his pill for the day. Otherwise, he takes

|        | 3 pills | 4 pills | 5 pills | 6 pills |
|--------|---------|---------|---------|---------|
| Type 1 | 003     | 013     | 113     | 114     |
| Type 2 | 030     | 031     | 131     | 141     |
| Type 3 | 012     | 022     | 023     | 123     |
| Type 4 | 300     | 301     | 311     | 411     |
| Type 5 | 102     | 202     | 203     | 213     |
| Type 6 | 120     | 220     | 230     | 231     |
| Type 7 | 111     | 211     | 221     | 222     |

a pill from the unique bin that results in leaving a configuration of the same type.

When the number of pills has dropped to three or fewer, it becomes difficult to maintain the type, but the pill taker can use the left-right rules for the types to decide how the configurations are further reduced. This amounts to using the following table:

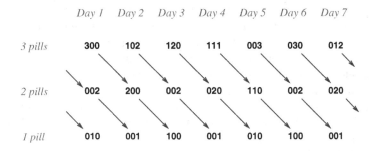

To use the table, the pill taker looks for the entry corresponding to his D and P; if it corresponds to reality, he should take the pill that produces the configuration below and to the right (following the diagonal line). Otherwise, the configuration will correspond to D+1 and he has already taken his pill for the day.

# Toughies

*Ask a difficult question, and the marvelous answer appears.*

> —*Molana Jalal-e-Din Mohammad Molavi Rumi,*
> *"Joy at Sudden Disappointment"*

The puzzles in this section are hard, but worth the effort. Several are variations or extensions of puzzles we have already considered.

The sample puzzle was posed as an open problem by Emil Kiss and K. A. Kearnes, in "Finite Algebras of Finite Complexity," *Discrete Math.*, Vol. 207 (1999), pp. 89–135. Petar Markovic brought the puzzle to a conference at MIT dedicated to the birthday of Professor Daniel J. Kleitman, in August 1999.

The problem was elegantly solved at the conference, by Noga Alon, Tom Bowman, Ron Holzman, and Danny himself. You're welcome, of course, to try this yourself first, but obviously you shouldn't feel bad if you don't see how to do it.

## Boxes and Sub-Boxes

Fix a positive number $n$. A *box* is a Cartesian product of $n$ finite sets; if the sets are $A_1, \ldots, A_n$, then the box consists of all sequences $(a_1, \ldots, a_n)$ such that $a_i \in A_i$ for each $i$.

A box $B = B_1 \times \cdots \times B_n$ is a *proper sub-box* of $A = A_1 \times \cdots \times A_n$ if $B_i$ is a proper nonempty subset of $A_i$ for each $i$.

Is it ever the case that a box can be partitioned into fewer than $2^n$ proper sub-boxes?

Solution: Partitioning into $2^n$ proper sub-boxes is easy as long as each $A_i$ has at least two elements. But no one at the conference could come up with an example with a partition into fewer than $2^n$ sub-boxes; in fact, it can't be done.

Consider first just one factor, $A_i$, and fix a proper nonempty subset $B_i \subset A_i$. Suppose we choose a *uniformly random* odd-size

119

subset $C_i \subset A_i$ (which may be all of $A_i$ if $|A_i|$ is odd). We claim that the probability that $|B_i \cap C_i|$ is odd is exactly $\frac{1}{2}$.

To see this, choose $C_i$ by running through the elements of $A_i$ one at a time, ending with an element of $B_i$ and an element of $A_i \setminus B_i$. We can decide by coin flip whether each element belongs to $C_i$ or not, except that the last decision is forced by the parity constraint on $|C_i|$. Then the penultimate coin flip will always determine the parity of $|B_i \cap C_i|$.

Of course, a sub-box $C = C_1 \times \cdots \times C_n$ of $A$ has odd size if and only if each $|C_i|$ is odd. It follows that if $B = B_1 \times \cdots \times B_n$ is a nonempty sub-box of $A = A_1 \times \cdots \times A_n$, and $C$ is a uniformly random odd-size sub-box of $A$, then the probability that $B \cap C$ has an odd number of elements is exactly $1/2^n$.

Now suppose we *do* have a partition of $A$ into fewer than $2^n$ sub-boxes, say $B(1), \ldots, B(m)$. Choose a uniformly random odd-size sub-box $C$ of $A$ as before, and note that with positive probability (at least $1 - m/2^n$), $C$ intersects every one of the sub-boxes $B(j)$ in an even number of elements.

But that's impossible, since $C$ itself has an odd number of elements. ♡

For the brave souls that are still with us, here are some more toughies. We begin with a puzzle that made *The New York Times*: "Why Mathematicians Now Care about their Hat Color," by Sara Robinson, April 10, 2001.

# Out-Guessing the Hat Colors

The hat-team is back.

This time, the color of each player's hat will be determined by a fair coin flip. The players will be arranged in a circle so that each player can see all the other players' hats, and no communication will be permitted. Each player will then be taken aside and given the option of trying to guess whether his own hat is red or blue, *but he may choose to pass.*

The outcome is drastic: Unless at least one player guesses, and every player who does guess guesses correctly, all the players will be executed. It sounds like the best plan must be to have only one player guess and the rest pass; that, at least, gives them a 50% chance of survival. But, incredibly, the team can do much better— with 16 players, for example, they can improve their odds to better than 90%. How?

If you think it's impossible to do better than 50%, it's a good sign that you understand the statement of the puzzle. But try it with three players before you give up.

The next puzzle's solution has a surprising connection, you will see, to the first's. For the rest of the puzzles, you're completely on your own.

## Fifteen Bits and a Spy

A spy's only chance to communicate with her control lies in the daily broadcast of 15 zeros and ones by a local radio station. She does not know how the bits are chosen, but each day she has the opportunity to *alter* any one bit, changing it from a 0 to a 1 or vice-versa.

How much daily information can she communicate?

## Angles in Space

Prove that among any set of more than $2^n$ points in $\mathbb{R}^n$, there are three that determine an obtuse angle.

## Two Monks on a Mountain

Remember the monk from Chapter 5 (Geometry), who climbed Mt. Fuji on Monday and descended on Tuesday? This time, he and a fellow monk climb the mountain on the same day, starting at the same time and altitude, but on different paths. The paths go up and down on the way to the summit (but never dip below the starting altitude); you are asked to prove that they can vary their speeds (sometimes going backwards) so that at *every* moment of the day they are at the same altitude!

## Controlling the Sums

Given a list of $n$ reals $x_1, \ldots, x_n$ from the unit interval, prove that you can find numbers $y_1, \ldots, y_n$ such that for any $k$, $|y_k| = x_k$ and

$$\left| \sum_{i=1}^{k} y_i - \sum_{i=k+1}^{n} y_i \right| \leq 2 \, .$$

# The Two-Bulb Room

Do you remember the prisoners and the room with one light switch? Again, each of $n$ prisoners will be sent alone into a room, infinitely often, but in some arbitrary order determined by their jailer. This time, however, there are *two* lights in the room, each with its own binary switch. There will be no means of communication other than these switches, whose initial states are not known. The prisoners again have a chance to confer in advance.

Again, we want to ensure that some prisoner will eventually be able to deduce that everyone has visited the room. What, you did it before with only *one* switch? Ah, but this time, every prisoner must follow the same set of rules.

# Area versus Diameter

Prove that among all closed regions of diameter 1 in the plane, the circular disk has the greatest area.

# The Even Split

Prove that from every set of $2n$ integers, you can choose a subset of size $n$ whose sum is divisible by $n$.

# Napkins in a Random Setting

Remember the conference banquet, where a bunch of mathematicians find themselves assigned to a big circular table? Again, on the table, between each pair of settings, is a coffee cup containing a cloth napkin. As each person sits down, he takes a napkin from his left or right; if both napkins are present, he chooses randomly.

This time there is no maitre d'; the seats are occupied in random order. If the number of mathematicians is large, what fraction of them (asymptotically) will end up without a napkin?

# Groups of Soldiers in the Field

Perhaps you also remember the soldiers in the field, each one watching the closest other one. Suppose there are a great many soldiers, in random positions in a large square, and they organize into the maximum possible number of groups subject to the condition that all watching goes on inside groups.

What will be the average size of a group?

## Ys in the Plane

You already know you can't fit uncountably many disjoint topological figure-8s in the plane. But you can certainly fit continuum-many line segments or circles. The next logical case is **Ys**: sets topologically equivalent to three line segments with a common end.

Can you prove that only countably many disjoint **Ys** can be drawn in the plane?

## More Magnetic Dollars

For our last toughie, we return to the Magnetic Dollars, but we strengthen their attractive power just a bit.

This time, an infinite sequence of coins will be tossed into the two urns. When one urn contains $x$ coins and the other $y$, the next coin will fall into the first urn with probability $x^{1.01}/(x^{1.01} + y^{1.01})$, otherwise into the second urn.

Prove that after some point, one of the urns will never get another coin.

# Solutions and Comments

### Guessing the Hat Colors

It is useful, as suggested, to try the game with three players first. At least you get to see how 50% can be improved upon. Generalizing from there is not, however, trivial.

With three players, each player is instructed to pass if the two hats he sees are of different colors, otherwise to guess that his own hat is the color he does *not* see. The result is that as long as both colors are represented (as in six of the eight possible configurations), the odd player will guess correctly and the other two will pass. Thus, the players win with probability 75%.

Notice that in the bad cases, where all three hats are the same color, *all* the players guess and they're all wrong. This is a crucial feature: The protocol packs six wrong guesses into only two configurations. Over all configurations, half the guesses must be wrong, thus the only way to win is to use the correct guesses efficiently and cram the wrong guesses together. Since our three-player protocol does these things as well as possible, it is optimal.

For $n$ players, we would ideally like to duplicate this feat and have just the two kinds of configurations: "good" ones where just one player guesses (correctly), and "bad" ones where everyone guesses and they are all wrong. This would force good configurations to outnumber bad ones by a factor of $n$, giving us a gratifying winning probability of $n/(n+1)$.

We have no chance of achieving this perfect optimum unless $n+1$ evenly divides the total number of configurations, $2^n$, which means that $n$ must itself be one less than some power of 2. Somewhat miraculously, this condition is sufficient as well as necessary.

The key is to find a set of bad configurations that has the property that every other configuration is adjacent to exactly one bad one (adjacent means you can get from one to the other by changing one hat color). Here is a way to define such a set.

Suppose $n = 2^k - 1$. Assign each player a different $k$-digit non-0 binary number (e.g., if there are 15 players, they get labeled 0001, 0010, 0011, ..., 1110, 1111.) These labels are treated as "nimbers,"[4] not numbers: You add them binarily without carry, so e.g., 1011 + 1101 = 0110 and anything plus itself is 0000.

The bad configurations are going to be the ones with the property that if you add all the labels of the red-hatted players, you get 0000. The strategy is this: Each player adds all the nimbers belonging to the people he sees whose hats are red. If the sum is 0000, he guesses that his own hat is also red. If the sum is his own nimber, he guesses that his own hat is blue. If the sum is anything else, he passes.

So why in the world should this work? Well, suppose the sum of the nimbers of *all* the people in red hats is, in fact, 0000. Then everyone with a blue hat on will compute 0000 as the red-hat sum, and will guess "red"; everyone with a red hat on will compute his own nimber as the sum, and will guess "blue." Thus, every player will guess and every one of them will be wrong—just what we want!

Now suppose that the sum of the red-hat nimbers is something else, say 0101. Then the *only* player who guesses is the one whose nimber is 0101, and his guess will be *right*.

---

[4]So called because of their usefulness in the game of Nim. As far as we know, the first appearance of this fortuitous terminology is on pg. 43 of *Winning Ways*.

The probability that the red-hat nimbers sum to 0000 is exactly 1/16 (as you might expect, since there are 16 possible sums). Thus, the strategy wins with probability 15/16; in general, with probability $1 - 2^{-k}$. It's worth checking that with nimbers of length 2, you get the three-player solution back.

If $n$ does not happen to be one short of a power of 2, the easiest thing for the players to do is to compute the largest $m < n$ which is of the form $2^k - 1$. Those $m$ players play as above and the rest refuse to guess regardless of what they see. At worst (if $n = 2^k - 2$ for some integer $k$), this results in a winning probability of $(n/2)/(n/2+1)$. These strategies are not generally best possible; for $n = 4$, you can't beat 75%, but for $n = 5$ (as pointed out to me by Elwyn Berlekamp), you can get $25/32 > 78\%$. Best strategy for general $n$ is a tough unsolved problem. ♡

The set of bad configurations we constructed above is not only a beautiful mathematical object, but one which is useful in real life. It is called a Hamming code and is an example of a perfect *error-correcting code*. Imagine that you are sending binary information over an unreliable channel which occasionally flips a bit. Group the bits you want to send into strings of size 11. There are $2^{15}/16 = 2^{11}$ red-blue sequences of length 15 with the property that the sum of the red-hat nimbers is 0000. These special strings, which you can write in binary (101010101010101 means every odd hat is red), are called "codewords"; since the number of codewords is $2^{11}$, you can associate one with every 11-bit binary string. One easy way to do this is to chop off the last 4 bits.

Now, instead of sending your 11-bit group, you send the unique associated 15-bit codeword. You pay a price in efficiency, but you get something back: reliability. This is because if one of the 15 bits is accidentally flipped, the person who gets your message can tell which one and flip it back!

How? When she receives the 15 bits, she can add up the red nimbers (those corresponding to ones in the sequence) and make sure they add up to 0000. Suppose they don't, e.g., they add up to 0101. So a bit must have been flipped; if it was only one bit, it must've been the fifth bit. So the receiver unflips the fifth bit, and checks her codebook to see which 11-bit sequence corresponds to the 15-bit codeword you intended to send. She will be right unless multiple bits were flipped.

The hat puzzle (in a somewhat different form) and solution were devised by Todd Ebert (now at the University of California, Irvine) in his 1998 PhD thesis at UC Santa Barbara. The puzzle came to my attention via Peter Gacs of Boston University. Interestingly, the Hamming code solution had been proposed some years earlier by Steven Rudich of Carnegie Mellon University for a related voting problem.

## 15 Bits and a Spy

Since there are 16 things the spy can do (change any bit or none), she can *in principle* communicate as much as four bits of information each day to her control. But how?

The answer is easy once you have the nimbers of the previous solution in your arsenal. The spy and her control assign the 4-digit nimber corresponding to the number $k$ to the $k$th bit of the broadcast, and their "message" is defined to be the sum of the nimbers of the ones in the broadcast.

The claim is that the spy can send any of the 16 possible messages at will, thus achieving a full four bits of communication. Suppose she wishes to send the nimber $n$ but the sum of the nimbers corresponding to the ones in the station's intended broadcast is $m \neq n$. Then she flips the $n+m$th bit. It makes no difference whether that bit was a 0 or a 1, since addition and subtraction of nimbers are the same thing. ♡

This puzzle came to me from Laci Lovász of Microsoft Research, who is uncertain of its origin.

## Angles in Space

I was tested on this puzzle during a visit to MIT, and was stumped. It makes sense that the $2^n$ corners of a hypercube represent the most points you can have in $n$-space without an obtuse angle. But how to prove it? Apparently, it was for some time an open problem of Paul Erdős and Victor Klee, then was solved by George Danzig and Branko Grünbaum.

Let $x_1, \ldots, , x_k$ be distinct points (vectors) in $\mathbb{R}^n$, and let $P$ be their convex closure. We may assume $P$ has volume 1 by reducing the dimension of the space to the dimension of $P$, then scaling appropriately; we may also assume $x_1$ is the origin (i.e., the 0 vector). If there are no obtuse angles among the points, then we claim that for each $i > 1$, the interior of the translate $P + x_i$ is disjoint from the interior of $P$; this is because the plane through $x_i$ perpendicular to the vector $x_i$ separates the two polytopes.

Furthermore, the interiors of $P + x_i$ and $P + x_j$, for $i \neq j$, are disjoint as well; this time via the separating plane running through $x_i + x_j$ perpendicular to the edge $x_j - x_i$ of $P$. We conclude that the volume of the union of the $P + x_i$, for $1 \leq i \leq k$, is $k$.

However, all these polytopes lie inside the doubled polytope $2P = P + P$, whose volume is $2^n$. Hence, $k \leq 2^n$ as claimed! ♡

## Two Monks on a Mountain

It is convenient to divide each path into monotone "segments" within which the path is always ascending, or always descending. (Level segments cause no problem as we can have one monk pause while the other traverses such a segment). Then we can assume that each such segment is just a straight ascent or descent, since we can have the monks modulate their rates so that their speed of ascent or descent is constant on any segment.

Label the $X$-axis on the plane by positions along the first monk's path, and the $Y$-axis by positions along the second monk's path. Plot all points where the two positions happen to be at the same altitude; this will include the origin (where both paths begin) and the summit (where they end, say at $(1,1)$). Our objective is to find a path along plotted points from $(0,0)$ to $(1,1)$; the monks can then follow this path, slowly enough to make sure that no monk is anywhere asked to move faster than he can.

Any two monotone segments—one from each path—which have some common altitude show up on the plot as a (closed) line segment, possibly of zero length. If we regard as a vertex any point on the plot which maps back to a segment endpoint (for either or both monks), the plot becomes a graph (in the combinatorial sense); and an easy checking of cases shows that except for the vertices at $(0,0)$ and $(1,1)$, all the vertices are incident to either 0, 2 or 4 edges.

Once we begin a walk on the graph at $(0,0)$, there is no place to get stuck or be forced to retrace but at $(1,1)$. Hence we can get to $(1,1)$, and any such route defines a successful strategy for the monks. ♡

The figure shows four possible landscapes, with one monk's path shown as a solid line, the other as dashed. Below each landscape is the corresponding graph. Note that, as in the last case, there may be detached portions of the graph which the monks cannot access (without breaking the common altitude rule).

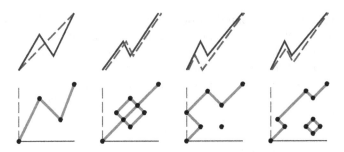

The puzzle was brought to my attention by Yuliy Baryshnikov of Bell Labs.

## Controlling the Sums

This puzzle arose "in real life," or, at least, in a serious mathematical problem involving optical networks: See A. Schrijver, P. D. Seymour and P. Winkler, "The Ring Loading Problem", *SIAM Review* Vol. 41, #4 (Dec. 1999), pp. 777–791. The authors believed the conjecture, but were embarrassed by being unable to prove it. The puzzle was made public, but still no proof or counterexample was offered; finally the author of this book found the proof below, which was pretty easy after all!

The problem is to "sign" a sequence of reals in [0,1] so as to control the sum even when an arbitrary final subsequence of the reals have their signs reversed. The natural first observation is that one can control all the *initial* sums by greedy signing, that is, putting $y_k = x_k$ when $\sum_{i=1}^{k-1} y_i \leq 0$ and $y_k = -x_k$ otherwise.

This insures that $|\sum_{i=1}^{k} y_i| \leq 1$ for all $k$, and rewriting gives

$$|\sum_{i=1}^{k} y_i - \sum_{i=k+1}^{n} y_i| = |2\sum_{i=1}^{k} y_i - \sum_{i=1}^{n} y_i|$$

$$\leq 2|\sum_{i=1}^{k} y_i| + |\sum_{i=1}^{n} y_i| \leq 3.$$

That's close, but unfortunately, a signing algorithm which does not look ahead can never get that 3 down to a 2. To see this, imagine that the sequence begins 1, .99, 1, .99, 1, .99, etc. for a hundred terms and then suddenly terminates with some number $z$.

The signs should alternate except at some point, and to know which point that is, you must know $z$.

However, note that the description above of the greedy signing requires a value for the "empty sum" in order to decide the first sign. Normally, we would say this sum is 0, but suppose instead we give it some real value $w$. Then, for fixed $w \in [-1, 1]$, the algorithm defines $y_k$ inductively by $y_k = x_k$ when $w + \sum_{i=1}^{k-1} y_i \leq 0$ and $y_k = -x_k$ otherwise. Then $w + \sum_{i=1}^{k} y_i \in [-1, 1]$ for all $k$; let $f(w) := w + \sum_{i=1}^{m} y_i$.

Suppose it happens that $f(w) = -w$. Then

$$\sum_{i=1}^{k} y_i - \sum_{i=k+1}^{m} y_i$$

$$= 2 \sum_{i=1}^{k} y_i - \sum_{i=1}^{m} y_i$$

$$= 2 \left( w + \sum_{i=1}^{k} y_i \right) - \left( w + \sum_{i=1}^{m} y_i \right) - w$$

$$= 2 \left( w + \sum_{i=1}^{k} y_i \right) \in [-2, 2]$$

as desired.

Since $f(-1) + (-1) \leq 0 \leq f(1) + 1$, the existence of a $w$ for which $f(w) = -w$ would follow from the intermediate value theorem, if $f$ were continuous. Of course, this is not the case; whenever a partial sum hits 0, some of the $y_i$ change sign and $f(w)$ may jump. (Since we have chosen to assign "+" when the partial sum is 0, $f$ will be continuous from the left.) However, it turns out that the *absolute value* of $f$ is continuous.

Note first that when no partial sum is at 0, the derivative $f'(w)$ is 1. On the other hand suppose $w = w_0$ is chosen such that one or more of the partial sums is zero; in particular, let $k \geq 0$ be minimal such that $w + \sum_{i=1}^{k} y_i = 0$. Then for sufficiently small $\varepsilon$, the signs of $y_j$ and $w + \sum_{i=1}^{j} y_i$, for $j > k$, flip as we move from $w = w_0$ to $w = w_0 + \varepsilon$. Hence, taking $j = m$, we have that $\lim_{w \to w_0^+} f(w) = -f(w_0)$.

It follows that when any partial sum hits zero, we will have $\lim_{w \to w_0^+} f(w) = -f(w_0)$; thus, the function $g$ given by

$g(w) = |f(w)|$ will be continuous everywhere, and differentiable except at finitely many points. The graph of $g$ is a zig-zag, with derivative 1 where $g(w) = f(w)$ and $-1$ where $g(w) = -f(w)$.

Of course, if we define $h$ by $h(w) = -w$, then the graph of $h$ is a line of slope $-1$ from $(-1, 1)$ to $(1, -1)$, which must intersect the graph of $g$. Moreover, it must either intersect at a point where $g'(w) = 1$ or coincide with a segment of the graph of $g$ of slope $-1$, in which case the left-most point of the segment also lies on the graph of $f$. Either way, we have a point $w$ at which $-w = g(w) = f(w)$. ♡

## The Two-Bulb Room

This puzzle is part of a serious problem in distributed computing, and the solution below, due to Steven Rudich of Carnegie Mellon University, is known as the "see-saw protocol." For more background, see M. J. Fischer, S. Moran, S. Rudich, and G. Taubenfeld, "The Wakeup Problem," *Proc. 22nd Symp. on the Theory of Computing*, Baltimore, Maryland, May 1990.

In the see-saw protocol, it is useful to think of one of the switches as the "pebble switch," which is occupied by a pebble or empty, and the other as the "see-saw" switch, which is left-down or right-down. Each prisoner begins with two virtual pebbles.

When first called upon to visit the room, a prisoner "gets on the see-saw" on the down side and pushes up. He stays on that side of the see-saw (that is, he remembers which side he got on) until he runs out of pebbles, in which case he lowers his side of the see-saw (this can only happen when he's on the up side) and dismounts, then leaves the playground and takes no further action.

While on the see-saw, he tries to give away a pebble whenever he's on the up side and take a pebble whenever he's down. To give away a pebble, he must find the pebble switch unoccupied; he then flips it and counts himself to have one fewer pebble. To take a pebble, he must find the pebble switch occupied, in which case he flips the switch and counts himself to have gained a pebble. If the pebble switch is not in an appropriate position, he does nothing.

When a prisoner collects $2n$ pebbles, he announces that everyone has visited the room. The conclusion is clear because there are $2n$ or $2n+1$ pebbles at the start (depending on the initial position of the pebble switch), and pebbles are not destroyed or created by the protocol, thus everyone must have contributed.

But why must we reach a state where one player has collected all the pebbles? Observe first that at *all* times between visits to the room, there are either (a) the same number of prisoners on both sides of the see-saw, or (b) one more on the up side. If (a) and someone gets on, he pushes up so we are now in (b); if someone gets off, he dismounts from the up side and pushes down, again producing (b). If (b) and someone gets on, he gets on the down side so equalizes to (a); if he dismounts, he reduces the number on the up side, again producing (a).

Now assume all prisoners have been in the room and that $k$ of them are presently on the see-saw (the others having run out of pebbles and dismounted). We know from the argument above that as long as $k > 1$, there will be at least one player on the up side of the see-saw and at least one on the down side. Then pebbles will flow from the up-prisoners to the down-prisoners until someone runs out, reducing the number on the see-saw to $k-1$. When $k$ drops all the way to 1, the remaining player will have all $2n$ or $2n+1$ pebbles and the protocol will end if it hasn't already.     ♡

So how does one invent a protocol like this? Beats me. Ask Rudich!

## Area versus Diameter

The origin of this puzzle and its lovely solution are not known to me; they have been around for at least 30 years, but maybe *much* longer than that. The solution does involve elementary calculus.

The diameter of a topologically closed, bounded region is the greatest distance between two points of the region. Note that it is *not* the case that every region of diameter 1 can be fit inside a circle of radius 1; for example, an equilateral triangle of side 1. No one knows the the area of the smallest region in which every region of diameter 1 can fit.

So, how do we show the disk has the largest area of all regions of diameter 1 if we can't fit the others inside? Let's let $\Omega$ be a closed region in the plane of diameter 1, and try to compute the area of $\Omega$ using polar coordinates. We can assume $\Omega$ is convex since taking its convex closure will not increase its diameter.

Let $P$ and $Q$ be points of $\Omega$ at distance 1, and place $\Omega$ on the plane so that $P$ is at the origin and $Q$ at $(1,0)$. Let $R(\theta)$ be the point of $\Omega$ farthest from $P$ in the direction $\theta$ (measured up from the $X$-axis), and let $r(\theta)$ be the distance from $P$ to $R(\theta)$. Then the area

$A$ of $\Omega$ is

$$\int_{-\pi}^{\pi} \frac{r(\theta)^2}{2} \, d\theta \, ,$$

which, since $r(\theta) \le 1$, is bounded by

$$\int_{-\pi}^{\pi} \frac{1}{2} \, d\theta = \frac{\pi}{2} \, .$$

This is twice the bound we were aiming for, but we shouldn't be too disappointed, since all we've done so far, in effect, is observe that $\Omega$ is contained in the right half of the disk of radius 1 centered at 0. We could cut this semidisc further, down to a lens shape, but how can we reduce the bound all the way to $\pi/4$?

The trick is to split the integral in two according to the sign of $\theta$, then change variables and recombine as follows:

$$\int_{-\pi}^{\pi} \frac{r(\theta)^2}{2} \, d\theta = \int_{-\pi}^{0} \frac{r(\theta)^2}{2} \, d\theta + \int_{0}^{\pi} \frac{r(\theta)^2}{2} \, d\theta$$

$$= \int_{0}^{\pi} \frac{r(\theta - \pi/2)^2}{2} \, d\theta + \int_{0}^{\pi} \frac{r(\theta)^2}{2} \, d\theta = \int_{0}^{\pi} \frac{r(\theta)^2 + r(\theta - \pi/2)^2}{2} \, d\theta \, ,$$

but $r(\theta)^2 + r(\theta - \pi/2)^2$ is the square of the distance between $R(\theta)$ and $R(\theta - \pi/2)$, by the Pythagorean Theorem. Thus, this expression is bounded by the square of the diameter of $\Omega$—namely, 1. Finally,

$$A \le \int_{0}^{\pi} \frac{1}{2} \, d\theta \le \frac{\pi}{4}$$

and we are done.                                                                                    ♡

### The Even Split

This puzzle, with $n$ replaced by 100, appeared in the 4th All Soviet Union Mathematical Competition, Simferopol, 1970. It is elegant enough to be called a theorem and in fact it is: in P. Erdős, A. Ginzburg, and A. Ziv, "Theorem in the Additive Number Theory," *Bull. Research Council of Israel*, Vol. 10F (1961), pp. 41–43. The proof below uses only elementary techniques.

Call a set "flat" if it sums to 0 modulo $n$. Let us note first that the statement we want to prove implies the following seemingly weaker statement: If $S$ is a flat set of $2n$ numbers, then $S$ can be split into two flat sets of size $n$. However, that in turn implies that any set of only $2n-1$ numbers contains a flat subset of size $n$ because we can add a $2n$th number to make the original set flat, then apply the previous statement to split this into *two* flat subsets of size $n$. One of these (the one without the new number) will do the trick.

So all three of these statements are equivalent. Suppose we can prove the second for $n = a$ and for $n = b$. Then if a set $S$ of size $2n = 2ab$ sums to 0 mod $ab$, it is, in particular, flat with respect to $a$, and we can peel off subsets $S_1, \ldots, S_{2b}$ of size $a$ which are also flat with respect to $a$. Each of these subsets $S_i$ has a sum we can write in the form $ab_i$. The numbers $b_i$ now constitute a set of size $2b$ which sums to 0 mod $b$, so we can split them into two sets of size $b$ which are flat with respect to $b$. The unions of the sets $S_i$ in each part are a bipartition of the original $S$ into sets of size $ab$ which are $ab$-flat, just what we wanted.

It follows that if we can prove the statement for $n = p$ prime, then we have it for all $n$. Let $S$ be a set of size $2p$, with the idea of creating a $p$-flat subset of size $p$.

How can we create such a subset? One natural possibility is to pair up the elements of $S$ and choose one element from each pair. Of course, if we do that, it will behoove us to ensure that the elements in each pair are different mod $p$, so our choice will not be of Hobson's variety. Can we do that?

Yes, order the elements of $S$ modulo $p$ (say, 0 through $p-1$) and consider the pairs $(x_i, x_{i+p})$ for $i = 1, 2, \ldots, p$. If $x_i$ were equivalent to $x_{i+p}$ mod $p$ for some $i$, then $x_i, x_{i+1}, \ldots, x_{i+p}$ would all be equivalent mod $p$ and we could take $p$ of them to make our desired subset.

Now that we have our pairs, we proceed by "dynamic programming." Let $A_k$ be the set of all sums (mod $p$) obtainable by adding one number from each of the first $k$ pairs. Then $|A_1| = 2$ and we claim $|A_{k+1}| \geq |A_k|$, and moreover, $|A_{k+1}| > |A_k|$ as long as $|A_k| \neq p$. This is because $A_{k+1} = (A_k + x_{k+1}) \cup (A_k + x_{k+1+p})$; thus if $|A_{k+1}| = |A_k|$, these two sets are identical, implying $A_k = A_k + (x_{k+1+p} - x_{k+1})$. This is impossible since $p$ is prime and $x_{k+1+p} - x_{k+1} \not\equiv 0 \mod p$, unless $|A_k| = 0$ or $p$.

Since there are $p$ pairs, we must eventually have $|A_k| = p$ for some $k \leq p$, hence $|A_p| = p$ and, in particular, $0 \in A_p$. The theorem follows. ♡

## Napkins in a Random Setting

We want to compute the probability that the diner in position 0 (modulo $n$) is deprived of a napkin. The limit of this quantity, as $n \to \infty$, is the desired limiting fraction of napkinless diners.

We may assume that everyone decides in advance whether to go for his right or left napkin, in case both are available; later, of course, some will have to change their minds or go without.

Say that diners $1, 2, \ldots, i-1$ choose "right" (away from 0) while $i$ chooses "left"; and diners $-1, -2, \ldots, -j+1$ choose "left" (again, away from 0) while diner $-j$ chooses "right."

If $k = i+j+1$, then the probability of this configuration is $2^{1-k}$. Note that $i$ and $j$ are at least 1, but, with high probability, less than $n/2$.

Observe that diner 0 loses out only when he is last to pick among $-j, \ldots, i$ and *none* of the diners $-j+1, \ldots, -2, -1; 1, 2, \ldots, i-1$ get the napkins they wanted. If $t(x)$ is the time at which diner $x$ makes his grab, then this happens exactly when $t(0)$ is the unique local maximum of $t$ in the range $[-j, -j+1, \ldots, 0, \ldots, i-1, i]$.

If $t$ is plotted on this interval, it looks like a mountain with $(0, t(0))$ on top; more precisely, $t(-j) < t(-j+1) < \cdots < t(-1) < t(0) > t(1) > t(2) > \cdots > t(i)$.

Instead of evaluating the probability of this event for fixed $i$ and $j$, it is convenient to lump all pairs $(i, j)$ together which satisfy $i + j + 1 = k$ for fixed $k$. Altogether there are $k!$ ways the values $t(-j), \ldots, t(i)$ can be ordered. If $T$ is the set of all $k$ grabbing times and $t_{\max}$ is the last of these, then each mountain-ordering is uniquely identified by the nontrivial subset of $T \setminus \{t_{\max}\}$ which constitutes the values $\{t(1), \ldots, t(i)\}$. Thus, the number of orderings that make a valid mountain is $2^{k-1} - 2$.

Finally, the total probability that diner 0 is deprived of his napkin is

$$\sum_{k=3}^{\infty} \frac{2^{1-k} \cdot 2^{k-1} - 2}{k!} = \left(2 - \sqrt{e}\right)^2 \approx 0.12339675. \heartsuit$$

Comparing this value to the fraction $9/64 = 0.140625$ achieved by the malicious maitre d', we see that he doesn't do all that much better than random.

Readers who would rather integrate than sum will prefer the following clean proof (simplified from an attack suggested by Aidan Sudbury of Monash University in Australia). We can assume that the "grabbing time" $t(i)$ for each diner is an independent, uniformly random real in the unit interval $[0, 1]$. Imagine that the diners form a doubly-infinite line, instead of a circle, and let $p(t)$ be the probability that a diner who grabs at time $t$ finds his right napkin already gone.

This occurs if his right-hand neighbor grabs first, and *either* chooses his left napkin voluntarily, or is forced to take the left napkin because *his* right napkin was already taken. Thus

$$p(t) = \frac{1}{2}t + \frac{1}{2}\int_0^t p(s)\,ds.$$

Differentiating with respect to $t$, then rearranging and integrating again, gives:

$$\frac{dp}{dt} = \frac{1}{2} + \frac{1}{2}p\,,$$

$$\frac{2}{1+p}\,dp = dt\,,$$

$$2\ln(1+p) = t + C\,,$$

but $C = 0$ since $p(0) = 0$. Hence,

$$p(t) = e^{t/2} - 1\,.$$

Of course, the probability that a diner grabbing at time $t$ finds his *left* napkin gone is the same, and here is the beauty of this approach: With $t$ fixed, the two events are independent. Hence the probability that our diner goes napkinless is $p(t)^2 = (e^{t/2} - 1)^2$ and averaging over grabbing times gives

$$\int_0^1 (e^{t/2} - 1)^2\,dt = (2 - \sqrt{e})^2.$$

$\heartsuit$

## Groups of Soldiers in the Field

Let us call two soldiers "mates" if they watch each other. As in the Insight chapter, in any group, the two soldiers closest to each other are mates, but there cannot be any other pair of mates in a group (of size, say $k$), because then the remaining $k-4$ "watchings" wouldn't be enough to hold the two pairs of mates and the $k-4$ singles together. So, if we can compute the probability $p$ that a given soldier has a mate, we could determine the average group size $g$: $p = 2/g$, thus $g = 2/p$.

Let us start with one soldier, $X$, in the middle of a square field $F$ of area 1 square mile. We then add $n$ more soldiers one at a time, each in a random position within $F$, where $n$ is huge. Let us call the second soldier $Y$, using lower case $x$ and $y$ to denote the positions of $X$ and $Y$. Let $N$ denote the event that $Y$ ends up as the nearest soldier to $X$, and $M$ the event that $Y$ ends up as $X$'s mate. Note that $\Pr(N) = 1/n$ since $Y$ is as likely as any subsequent soldier to be the closest to $X$. We want to calculate $p = \Pr M / \Pr N$.

For $N$ to occur, we require that no subsequent soldiers arrive inside the circle through $y$ with center at $x$. For $M$ to occur, there must be no subsequent arrivals inside either in this circle or in the overlapping one through $x$ with center at $y$. The ratio of the first area to the second is $c := \pi/(\frac{4}{3}\pi + \frac{\sqrt{3}}{2}) \approx 0.6215049$. (Of course, this ratio does not depend on the distance $r$ from $x$ to $y$; the figure below indicates how to calculate $c$ using unit disks.)

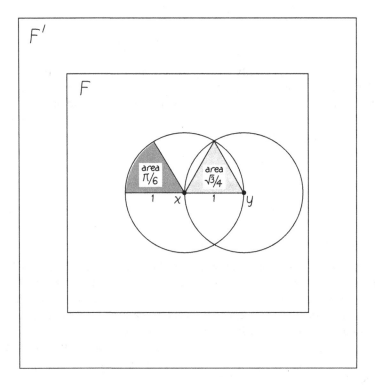

Suppose we mark a field $F'$ which contains $F$, but has area $1/c$ square miles. Let $M'$ be the event that, *if the rest of the soldiers are placed randomly in $F'$ instead of in $F$*, $Y$ ends up as the mate to $X$. Regardless of the value of $r$, each new soldier in $F'$ has the same probability of ruining $M'$ as the new soldiers in $F$ had of ruining $N$, so $\Pr M' = \Pr N = 1/n$.

Now suppose $Y$ itself is chosen from all of $F'$ instead of just $F$. To have a chance to be $X$'s mate, it must be in the smaller field, which will happen with probability $c$; and we have seen that if it *does* land in $F$, it will end up as $X$'s mate with probability $1/n$. Thus, altogether, $Y$ has probability $c/n$ so $p = c$.

It follows that the mean size of a group is $2/p \approx 3.2179956$. ♡

The above reasoning is not completely rigorous since the issues of limits and edge effects are not addressed. Fans of calculus and of Poisson point distributions will find it more straightforward and perhaps more convincing as well to calculate $p$ by integrating over $r$, leading to

$$p = \int_0^\infty e^{-\pi r^2/c} 2\pi r\, dr.$$

However, the above scaling argument is more general as well as more elementary, and except for the calculation of $c$, is independent of dimension. If the soldiers are on a line, the ratio $c$ is $2/3$ giving an average group size of 3; in space (frogmen, perhaps?), $c = 16/27$, giving a mean group size of $3\frac{3}{8}$. As the dimension increases, $c \to 1/2$ so $g \to 4$. Curious, isn't it, that the answers are rational in dimensions 1 and 3, but not on the plane?

Luis Goddyn of Simon Fraser University, who brought this nice problem (with its calculus solution) to my attention, points out that it would be equally interesting to know the probability that a given soldier is not being watched. Neither he nor I knows how to calculate that number, which he believes empirically to be about 28% on the plane (it's 25% on the line). Incidentally, the graph defined on a metric space by connecting each point to its closest neighbor is often called the Gabriel graph.

## Ys in the Plane

Here is a neat proof supplied by Randy Dougherty of Ohio State University. Associate with each **Y** three rational circles (rational center and radius) containing the endpoints, and small enough so that none contains or intersects any other arm of the **Y**. We claim that no *three* **Y**s can all have the same triple of circles; for, if that were so, you could connect the hub of each **Y** to the center of each circle by following the appropriate arm until you hit the circle, then following a radius to the circle's center. This would give a planar embedding of the graph $K_{3,3}$, sometimes known as the "gas-water-electricity network."

In other words, we have created six points in the plane, divided into two sets of three each, with each point of one set connected by a curve to each point of the other set, and no two curves crossing. This is impossible; in fact, readers who know Kuratowski's Theorem will recognize this graph as one of the two basic nonplanar graphs.

To see for yourself that $K_{3,3}$ cannot be embedded in the plane without crossings, let the two vertex sets be $\{u, v, w\}$ and $\{x, y, z\}$. If we could embed it without crossings the sequence $u, x, v, y, w, z$ would represent consecutive vertices of a (topological) hexagon. The edge $uy$ would have to lie inside or outside the hexagon (let us say inside); then $vz$ would have to lie outside to avoid crossing $uy$, and $wx$ has no place to go.                                          ♡

## More Magnetic Dollars

This variation of Polya's urn problem was studied by Joel Spencer of NYU and his student Roberto Oliveira. The really neat way to show that one urn gets all but finitely many coins is to employ those memoryless waiting times that proved so useful in Version II of the Gladiators problem, from the *Games* chapter.

Look just at the first urn and suppose that it acquires coins by waiting an average of $1/n^{1.01}$ hours between the $n$th coin and the $(n+1)$st coin, where the waiting time is memoryless. Coins will arrive slowly and sporadically at first, then faster and faster; since the series $\sum_{n=1}^{\infty} 1/n^{1.01}$ converges, the urn will explode with infinitely many coins at some random moment (averaging about 4 days, 4 hours, and 35 minutes after the process begins).

Now suppose we start two such processes simultaneously, one with each urn. If at some time $t$, there are $x$ coins in the first urn and $y$ in the second, then (as we saw with the gladiator-light bulbs) the probability that the next coin goes to the first urn is

$$\frac{1/y^{1.01}}{1/x^{1.01} + 1/y^{1.01}} = \frac{x^{1.01}}{x^{1.01} + y^{1.01}} \, ,$$

exactly what it should be. Nor does it matter how long it's been since the $x$th coin in the first urn (or $y$th in the second) arrived, since the process is memoryless. It follows that this accelerated experiment is faithful to the puzzle.

However, you can see what happens now; with probability 1, the two explosion times are different. (For this, you only need to know that the first waiting time has a continuous distribution.) But the experiment ends at the first explosion, at which time the other urn is stuck with whatever finite number of coins it had. ♡

Seems like kind of a scary experiment, doesn't it? The slow urn never got to finish because, in effect, time ended.

# Unsolved Puzzles

Man can learn nothing unless he proceeds from the known to the unknown.
—*Claude Bernard (1813–1878)*

To quote one of my friends: "What the ?$%&#@! is an unsolved puzzle?"

Indeed, it's impossible to know if there is an elegant solution if no solution at all has been found. Still, some unsolved puzzles attract quite a lot of attention, owing to elegance of the problem and, often, amazement that no solution is known.

Mathematicians, especially those—like your author—brought up in the Erdösian tradition of looking for the simplest thing you don't know, often brag about such problems. Put several such fanatics together, and you often hear a conversation something like:

"Here's something that's been bothering me; do you know the answer?"

"Actually, I'm not sure I even know the answer to this simpler question."

"Are you kidding? *I* don't even know *this!*"

Of course, we must distinguish between an unsolved puzzle and an unsolved *problem*, like the Riemann Hypothesis or the question of whether P=NP. Unsolved problems may or may not be elegant and elementary to state; but they are important, and are studied because they arise (often as obstacles) in the pursuit of mathematics. The statement of an unsolved problem often requires "professional" mathematical concepts (graphs, groups, manifolds, transformations, representations, etc.) which we do not permit in a puzzle—although they may be implicit in the statement or, ultimately, necessary in the solution.

Unsolved puzzles should be entertaining, intriguing, even galling. But they should not be important *as far as we know*. Of course, every such puzzle has a certain unavoidable level of importance because it represents a gap in our mathematical arsenal. A solution to an

141

unsolved puzzle might unveil a valuable, serious technique; or it might result from the application of some very deep mathematics far beyond the scope of this book. Some puzzles below, like the Union-Closed Set Conjecture and the $3x + 1$ Conundrum, have attracted so much attention in the mathematical community that one could reasonably argue that *any* solution would be of huge interest regardless of its applicability elsewhere.

These puzzles are presented here for your amusement and to remind us all of how little we know. If even one of them is solved by someone who learned of it from this book, it would be a minor miracle. If you *think* you've solved one, you're probably wrong. Use the references below, your professional mathematical friends, and your favorite web search engine to learn more about other attempts to solve the puzzle. With luck, you will find out which well-known trap you fell into before you embarrass yourself in public.

If you still think you have a valid solution, it should be written up and submitted to a suitable mathematics journal. Don't send it to me: I am not an expert on *any* of the problems.

In this chapter, there will of course be no solutions section, but we will continue the format of puzzles first and then comments and source information. We begin with a classic from John H. Conway. Good luck!

## Conway's Angel and Devil

An Angel flies over an infinite checkerboard, and every now and then she must alight on a square. She can travel no more than 1000 King-moves in the air before she lands.

While she's in the air, however, the Devil, who lives below the board, can destroy any one square of his choice.

Can the Devil trap the Angel?

## The 3x+1 Conundrum

Beginning with an arbitrary positive integer, repeat: If the number is even, halve it; if odd, treble it and add 1.

Prove that you will eventually cycle; even better, that you will eventually enter the cycle 1, 4, 2, 1, 4, 2 ....

# Longest Common Subsequence

Two random binary sequences of length $n$ are generated, with each entry being "1" independently with probability $p$. Let $C_p(n)$ be the length of the longest common subsequence of the two sequences, and let $C_p$ be the limit of the ratio $C_p(n)/n$.

Compute $C_{\frac{1}{2}}$—or, at least, prove that $C_{\frac{1}{2}} < C_p$ for $p \neq \frac{1}{2}$.

# Squaring the Lake

Prove that every simple closed curve in the plane contains four points forming the vertices of a square.

# The Lonely Runner

In an unending race, $n$ runners having distinct constant speeds start at a common point and run laps on a unit length circular track. Prove that each runner will at some moment in time be at distance at least $1/n$ from every other runner.

# Sorting Pairs in Bins

Standing in a row are $n$ bins, the $i$th bin containing two balls numbered $n+1-i$. At any time, you may swap two balls from adjacent bins. How many swaps are needed to get every ball into the bin carrying its number?

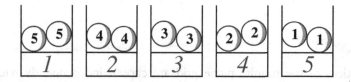

# Unfolding the Polyhedron

Prove that it is always possible to cut a convex polyhedron's edges so its surface unfolds into a simple planar polygon.

## Illuminating the Polygon

Is every polygonal region in the plane, with reflecting edges, illuminable from some interior point?

## Conway's Thrackles

A *thrackle* is a drawing on the plane consisting of vertices (points) and edges (non-self-intersecting curves) such that:

- every edge ends at two different vertices, but hits no other vertex; and

- every edge intersects each other edge exactly once, either at a vertex or by crossing at an interior point.

Is there a thrackle with more edges than vertices?

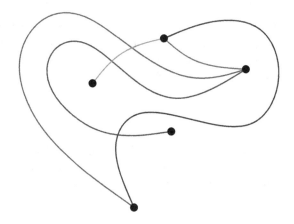

## Gridlock

Vertices of the infinite plane grid are chosen independently with probability $p \in (0, 1)$, and according to fair coin flips, each chosen vertex is occupied either by a car facing north or a car facing east.

The cars are controlled by a traffic signal which alternates "green-east" and "green-north." When it turns green-east, each eastbound car whose right-hand neighbor vertex is unoccupied moves to that vertex; the others (including those blocked by another eastbound car) remain where they are.

When the signal turns green-north, each unblocked northbound car moves one vertex northward.

Experiments suggest that when $p$ is below a certain critical value $p_0$, the cars gradually break free; that is, each has a limiting velocity equal to the velocity of a car which is never blocked. But when $p > p_0$, the opposite occurs: The cars get hopelessly tangled and every car takes only finitely many moves before being blocked forever.

Your mission, if you choose to accept it, is to prove *any* of this!

## The Midlevels Conjecture

Prove that you can cycle through all the subsets of size $n$ or $n+1$ of a set of size $2n+1$ by adding or deleting one element at a time.

## Building Venn Diagrams

An $n$-Venn diagram is a collection of $n$ simple closed curves in the plane, all of whose intersections are simple crossings, having the property that for any subset of the curves, the set of points inside the curves of the subset, and outside the other curves, is a nonempty connected component of the plane minus the union of the curves.

Can every $n$-Venn diagram in the plane be extended to an $n+1$-Venn diagram?

## A Strategy for Chomp

A number $k$ is fixed, and Alice and Bob play the following game. Alice names a divisor of $k$. Bob names another divisor of $k$, but it must not be a multiple of Alice's last call. Alice names a third divisor that is not a multiple of any previous call, etc.; the loser is the one who names "1."

Note that this game generalizes the Chomp game from a previous chapter; $k = 2^{m-1}3^{n-1}$ is equivalent to playing on an $m \times n$ chocolate bar. The earlier proof generalizes as well, but we are left with the following puzzle, both for the chocolate bar version and its generalization:

Find a winning strategy for Alice!

# All Roads Lead to Rome

Suppose that a network (not necessarily planar) of cities and one-way roads has the following properties: From each city, there are exactly two roads leading out, and for some $n$, you can get from any city to any other city in $n$ steps.

Prove that you can color the roads red and blue in such a way that (a) each city has an exit road of each color, and (b) there is a set of instructions (e.g., "*RBBRRRBRBBR*") that always ends at the same city, regardless of your beginning point.

# Discs in a Disc

Prove that any set of discs of total area 1 can be packed into a disk of area 2. Even better, prove that in $d$-dimensional space, any set of copies of a convex figure, with total volume 1, can be packed into a copy of volume $2^{d-1}$.

# The Union-Closed Set Conjecture

Let $U$ be a finite set and $\mathcal{F}$ a family of nonempty subsets of $U$ which is closed under unions. Prove that there is an element of $U$ which is in at least half the sets of $\mathcal{F}$.

# Comments and Sources

## Conway's Angel and Devil

A recent progress report on this fascinating puzzle can be found in the very nice collection, *Games of No Chance*, Richard J. Nowakowski, editor, Cambridge University Press, Cambridge, 1996. (Or see http://www.msri.org/publications/books/Book29/files/conway.pdf.)

Elwyn Berlekamp has shown that if the Angel has "power 1," that is, can only make single King's moves, the Devil can win. It may be that, in fact, the Devil can win no matter what the Angel's power is; however, as far as we know, power 2 is enough for the Angel to survive forever.

It seems the Angel of power 1000 ought to be able to win, but, as the puzzle's inventor John H. Conway points out in his article, there are several annoying features that seem to get in the way of a solution. One issue is that the Devil can never make a mistake; no matter what squares he destroys, he'll always be better off than he was in the initial configuration. Another issue is that he seems to have an answer to any reasonable "potential function" strategy that the Angel may have that tells her where to go depending on what squares have been eaten.

Furthermore, if the Angel has some seemingly mild handicap like not being permitted to travel to a point more than $10^{99}$ squares south of a square she has already visited, the Devil can win.

Conway himself still believes in the Angel, as reflected in the fact that he offers a prize of $1,000 for a proof that the Devil can win, but only $100 for a winning strategy for some sufficiently high-powered Angel.

## The 3x+1 Conundrum

The origins of this famous puzzle, also known as the Collatz problem, the Syracuse problem, Kakutani's problem, Hasse's algo-rithm, and Ulam's problem, are obscure. A student at the University of Hamburg named Lothar Collatz entered a similar problem into his notebook on July 1, 1932, but the problem as it is known now seems to have reached popularity only in the 1950s.

Jeff Lagarias of AT&T Labs has written a very nice survey, "The $3x+1$ Problem and its Generalizations," in the *Amer. Math. Monthly*, Vol. 92 (1985), pp. 3–23. This paper is available on the web at http://www.cecm.sfu.ca/organics/papers, and more information can be found at Lagarias' own website, http://www.research.att.com/~jcl/3x+1.html.

Lagarias points out that at one time the puzzle was said to be part of a conspiracy to slow down mathematical research in the US; let that be a warning!

## Longest Common Subsequence

This puzzle goes back at least 30 years, to the 1974 PhD Thesis of V. Dančík at the University of Warwick. Michael Steele (University of Pennsylvania) conjectured that $C_{1/2} = 2/(1 + \sqrt{2}) \approx 0.828427$. V. Chvátal and D. Sankoff showed that $0.773911 < C_{1/2} < 0.837623$, and it began to look like Steele's number was too high; finally George Lueker (UC Irvine) killed the conjecture in 2003, getting $0.7880 < C_{1/2} < 0.8263$. A short report appears in

*Proc. 14th ACM-SIAM Symp. on Discrete Algorithms*, Baltimore, Maryland, 2003, pp. 130–131.

Proving that $C_p$ exists is an easy exercise in subadditivity (see, e.g., Rick Durrett's *Probability: Theory and Examples*, Wadsworth, 1991, Section 6.6), but the method often leaves you with no clue as to how to compute the constant. Another such example: Béla Bollobás and I showed that there is a number $K_d$ with the property that the longest coordinate-wise increasing chain among $n$ random points in $d$-space has size about $K_d \cdot n^{1/d}$. We know $K_1 = 1$, $K_2 = 2$, and $\lim_{d \to \infty} K_d = e$, but what is $K_3$?

If we vary the probability $p$ of a "1," letting $p > 1/2$, we of course get $C_p > p$ since we can look at the subsequences consisting of all ones. Thus, $C_p \to 1$ as $p \to 1$, and it stands to reason that $C_p$ is minimized at $p = 1/2$. We shouldn't even have to know any exact values of $C_p$ to prove it. But, at the moment, no one knows how.

## Squaring the Lake

At http://www.ics.uci.edu/~eppstein/junkyard/jordan-square. html you can find a nice discussion of this puzzle. It seems that there have been some proofs that *sufficiently smooth* closed curves in the plane always contain the corners of a square; for example, Walter Stromquist, "Inscribed Squares and Square-Like Quadrilaterals in Closed Curves," *Mathematika*, Vol. 36 (1989), No. 2, pp. 187–197. The general conjecture has remained open, however, for more than 90 years. See *Old and New Unsolved Problems in Plane Geometry and Number Theory*, by Stan Wagon and Victor Klee, Mathematical Association of America, 1991.

It's a little embarrassing that mathematicians cannot determine whether every closed curve in the plane contains the corners of a square, don't you think?

## The Lonely Runner

This lovely conjecture seems to be due originally to J. M. Wills, in a paper entitled "Zwei Satze uber Inhomogene Diophantische Approximation von Irrationalzahlen, *Monatsch. Math.*, Vol. 71 (1967), pp. 263–269. In 1973 it was independently arrived at by T. W. Cusick, who teamed with Carla Pomerance in 1984 to prove it for up to five runners. Tom Bohman, Ron Holzman and Dan Kleitman (remember them from Boxes and Sub-Boxes?) have got it up to six; you can see their paper at http://www.combinatorics.org/Volume_8 /PDF/v8i2r3.pdf. At http://www.ceremade.dauphine.fr/CMD/ preprints03/0315-6runnersJT.pdf you can find a shorter proof by Jérôme Renault.

The puzzle's name is due to Luis Goddyn of Simon Fraser University, who has also contributed to the literature.

The puzzle is number-theoretic in character; in fact, it can be shown that one can assume all the speeds are integers.

## Sorting Pairs in Bins

This curious puzzle arose at Bellcore (now Telcordia Technologies) in connection with statistical surveys involving preference orders with ties. I worked on it with colleagues Michael Littman (now of Rutgers) and Graham Brightwell (London School of Economics). The puzzle generalizes not only to $k$ balls in a bin, but to bins of different sizes. We will concentrate here on bins of size 2.

If we just had, say, $n$ bins of size 1 containing $n$ balls numbered in reverse order, then it's a standard and easy exercise to work out that $\binom{n}{2}$ swaps are needed to get each ball into the right bin. One way to see this is to observe that every pair of balls begins in the wrong order, and an adjacent swap fixes only one pair. This also tells us that as long as we don't do anything stupid (namely, swapping two balls that are already in the desired order), we will be finished after $\binom{n}{2}$ swaps. In fact, no matter what the initial configuration, $\binom{n}{2}$ swaps suffice; the reverse order that we began with is, as you might expect, the worst case.

It looks obvious that with two balls to a bin, the same arguments work. You can imagine that the balls come in two sets, red and green, each numbered 1 through $n$; we can separately sort each set to finish in $2\binom{n}{2}$ steps. And surely $2\binom{n}{2}$ are necessary, right?

Well ... no. Take $n = 5$ and look at the diagram below; it appears that we have magically sorted the balls in only 15 swaps, instead of the 20 that seemed unavoidable.

One cannot do better than 15 swaps, or more generally, $\lceil \binom{2n}{2}/3 \rceil$ swaps when there are $n$ two-ball bins. To see this, award 1 point for a pass, i.e., a high-numbered ball passing from left of a low-numbered one to the right. We assign $1/2$ point each for catching up and moving on if the passing is done in two stages. In addition, two

| 1 | 2 | 3 | 4 | 5 |
|---|---|---|---|---|
| 55 ←→ 44 | 33 | 22 | 11 | |
| 54 | 54 ←→ 33 | 22 | 11 | |
| 54 | 43 | 53 ←→ 22 | 11 | |
| 54 | 43 | 32 | 52 ←→ 11 | |
| 54 | 43 | 32 ←→ 21 | 51 | |
| 54 | 43 ←→ 21 | 32 | 51 | |
| 54 ←→ 31 | 42 | 32 | 51 | |
| 41 | 53 ←→ 42 | 32 | 51 | |
| 41 | 32 | 54 ←→ 32 | 51 | |
| 41 | 32 | 42 | 53 ←→ 51 | |
| 41 | 32 | 42 ←→ 31 | 55 | |
| 41 | 32 ←→ 21 | 43 | 55 | |
| 41 ←→ 21 | 32 | 43 | 55 | |
| 11 | 42 ←→ 32 | 43 | 55 | |
| 11 | 22 | 43 ←→ 43 | 55 | |
| 11 | 22 | 33 | 44 | 55 |

counters with the same number must incur a 1-point charge because they must separate (1/2 point) at some point and then recombine. Therefore, there are $\binom{2n}{2}$ points that must be collected in the process of sorting all the balls.

In one step, how many points can be taken care of? Well, suppose balls numbered $u$ and $y$ are exchanged between a bin containing $u$ and $v$ and an adjacent bin containing $x$ and $y$. We can get 1 point for $u$ versus $y$, $1/2$ each for $u$ versus $v$ (for "moving on") and $y$ versus $x$, and $1/2$ each for $u$ versus $x$ (for "catching up") and $y$ versus $v$, for a total of 3 points. The bound follows.

There's more good news. As in the case of bins of size 1, it is not hard to show that reversing the order is again the hardest case; so if $f_2(n)$ is the minimum number of swaps needed to sort $n$ bins and two balls to a bin, from any initial configuration, then $f_2(n)$ is necessary for this puzzle. It's also easy to prove that if a swap is to be made between bins $i$ and $i+1$, it cannot be wrong to swap the highest numbered ball in bin $i$ with the lowest numbered ball in bin $i+1$.

But there's bad news, or the puzzle wouldn't be in this chapter. The bound $\lceil \binom{2n}{2}/3 \rceil$ is not always achievable; for example, it shows that $f_2(6) \geq 21$, but in fact a computer search found no way to do the six-bin case in fewer than 22 swaps. Worse, the nice-looking swap pattern seen in the diagram for five bins is not generally optimal.

But it's perfectly possible that some other scheme *is* optimal, and perhaps even provides a nice formula for $f_2(n)$.

## Unfolding the Polyhedron

Arguably, this puzzle is *really* old: See *The Painter's Manual: A Manual of Measurement of Lines, Areas, and Solids by Means of Compass and Ruler Assembled by Albrecht Dürer for the Use of All Lovers of Art with Appropriate Illustrations Arranged to be Printed in the Year MDXXV*, reprinted by Abaris Books in 1977. The theory is that if you want to decorate a polyhedron, it is useful to imagine it cut along its edges and laid out on the plane without overlapping.

The specific statement of the puzzle seems, however, to have originated in a paper by G. C. Shephard of the University of East Anglia: "Convex Polytopes with Convex Nets," *Math. Proc. Camb. Phil. Soc.*, Vol. 78 (1975), pp. 389–403.

It is known that there are *non*convex polytopes which cannot be cut and laid out in this manner; and there are convex polytopes which have self-overlapping unfoldings as well as proper ones. Pictured below is an example involving a mere tetrahedron, supplied by Makoto Namiki of the University of Tokyo.

Incidentally, for those who may be wondering, it is not the case that every unfolding folds back up to a convex polyhedron in an unambiguous manner.

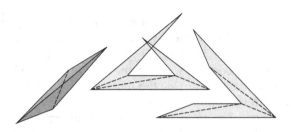

For a discussion and some additional pictures, the reader is referred to the homepage of Komei Fukuda at ETH Zurich, http://www.ifor.math.ethz.ch/~fukuda/fukuda.html.

## Illuminating the Polygon

According to Joseph O'Rourke of Smith College, in his book *Art Gallery Theorems and Algorithms* (Oxford University Press, 1987), the puzzle's original poser is unknown. Victor Klee wrote about in in 1969, in an article in the *Americal Mathematical Monthly* which attracted much attention.

If, as originally stipulated, a light ray hitting a vertex is absorbed, it is possible to construct a polygon which is not illuminable from *some* interior point. The example illustrated below was found by G. Tokarsky in 1995.

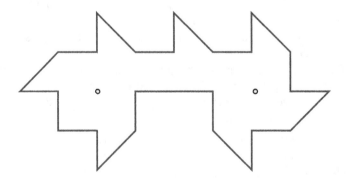

O'Rourke conjectures that in any mirror polygon $P$ the set of interior points from which $P$ cannot be illuminated has measure 0; moreover, if vertices are replaced by small circular arcs, there are no such points at all.

It is possible to design a *curved* closed figure that cannot be illuminated from *any* interior point, as discovered by Klee himself. His figure (shown here) makes use of two half-ellipses with foci at the indicated points; a light source in the top half, for example, leaves the left and right lobes on the lower half dark.

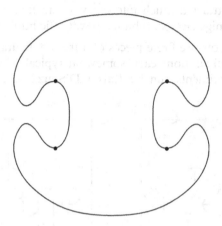

There are quite a few other intriguing open puzzles involving mirrors. For example, can a finite set of disjoint segment-mirrors trap the light from a source? How about circle mirrors? These and more can be found in slides from a wonderful talk by O'Rourke, "Unsolved Problems in Visibility," http://dimacs.rutgers.edu/dci/2001/Visibility.ppt.

## Conway's Thrackles

This intriguing conjecture of Conway's dates from the '60s; see D. R. Woodall, "Thrackles and Deadlock," in *Combinatorial Mathematics and its Applications*, Proceedings of a Conference held at the Mathematical Institute (D. A. J. Welsh, editor), Oxford (1969), pp. 335–348. To make the puzzle's status even more embarrassing, the conjecture reduces merely to establishing that the union of two even cycles which share a point can never be drawn as a thrackle. The best partial result I know of is that the number of edges cannot exceed *twice* the number of vertices minus 3 (L. Lovász, J. Pach, and M. Szegedy, "On Conway's Thrackle Conjecture," *Discrete and Computational Geometry*, Vol. 18 (1997), pp. 369–376).

There is a thrackle fan club centered at http://www.thrackle.org.

## Gridlock

This model of traffic flow at the intersection of two major one-way streets was introduced in O. Biham, A. A. Middleton, and D. Levine, "Self Organization and a Dynamical Transition in Traffic Flow Models", *Phys. Rev. A*, Vol. 46:R6124 (1992). Its bizarre behavior has attracted much interest; you can find a bibliography at http://cui.unige.ch/spc/Bibliography/traffic.html.

Pictured below are finite pieces of a free configuration and of a gridlocked configuration, each somewhat typical of what has appeared in experiments run by Raissa D'Souza of Microsoft Research.

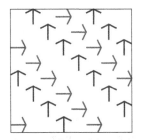 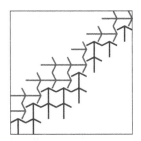

Now, if we could just prove that for *some p*, even very near 0 or very near 1, things actually behave like that...

## The Midlevels Conjecture

This famous Hamilton cycle puzzle has been attributed at various times to combinatorialists Ivan Hável, Claude Berge, Italo Dejter, Paul Erdös, W. T. Trotter, and David Kelley; Hável may have been the first. Of course, the question is natural enough to have been rediscovered often. Kelley presented the puzzle at a meeting in Oberwolfach, Germany in 1981 and was given a prize (a bottle of wine) for shortest problem presentation.

Warning to the reader: This puzzle is infectious, and indeed experimentation has led many intelligent investigators to believe that there is a pattern that will work for any $n$. No one believes the conjecture is false; indeed Robert Roth (Emory University) ran some computer experiments years ago that suggest that the *number* of ways to cycle through the middle levels is an *extremely* rapidly increasing function of $n$. That it should drop to 0 for some $n$ seems highly implausible, but no proof to the contrary exists.

The best partial result can be found in the recent PhD thesis of Robert Johnson, a student of Imre Leader of Cambridge University. Johnson showed that as $n$ increases, there are cycles through an arbitrarily high fraction of the midlevel sets.

## Building Venn Diagrams

This puzzle, like the last one, is about Hamilton cycles; to add a new area to an $n$-Venn diagram, you need to draw a closed curve which crosses each area exactly once. The conjecture is actually due to your author, and found in "Venn Diagrams: Some Observations and an Open Problem," *Congressus Numerantium*, Vol. 45 (1984), pp. 267–274.

In Kiran B. Chilakamarri, Peter Hamburger, and Raymond E. Pippert, "Simple, Reducible Venn Diagrams on Five Curves and Hamiltonian Cycles," *Geometriae Dedicata*, Vol. 68 (1997), pp. 245–262, the authors prove that if crossings of more than two curves are permitted, you can indeed extend any Venn diagram. But the original conjecture remains open, now for 20 years.

The *Electronic Journal of Combinatorics* runs some useful web surveys, among them one on Venn diagrams by Frank Ruskey of the University of Victoria. You can find it at http://www.combinatorics. org/Surveys/ds5/VennEJC.html. Ruskey is an expert on Venn diagrams, which go back to John Venn's article, "On the Diagrammatic and Mechanical Representation of Propositions and Reasonings," in *The London, Edinburgh, and Dublin Philosophical Magazine and Journal of Science*, Vol. 9 (1880), pp. 1–18.

Being around for 123 years does not make a subject immune to elementary new discoveries, for sure! Recently, a different Venn diagram problem was solved when Chip Killian and Carl Savage of North Carolina State University, and Jerry Griggs of the University of South Carolina, showed how to make a rotationally symmetric Venn diagram of any prime order. An article by Barry Cipra about their work can be found at http://www.math.ncsu.edu/News/venn.pdf.

## A Strategy for Chomp

Chomp was invented by David Gale in 1974 ("A Curious Nim-Type Game," *Amer. Math. Monthly* Vol. 81 (1974), pp. 8760–879) and christened by Martin Gardner. However, it is equivalent to a game called "Divisors," published in Fred. Schuh, "Spel van Del-ers," *Nieuw Tijdschrift voor Wiskunde,* Vol. 39 (1952), pp. 299–304. In Shuh's game, a positive integer $n$ is fixed and the players take turns naming divisors of $n$, under the constraint that no play may be a multiple of a previous play; the loser is the one forced to call "1."

If $n$ is of the form $p^a q^b$ where $p$ and $q$ are prime, then all plays are of the form $p^i q^j$ for $0 \leq i \leq a$, $0 \leq j \leq b$, and each play must have an $i$ or a $j$ which is smaller than any previously used $i$ or $j$. But this is the same as playing Chomp with an $(a+1) \times (b+1)$ chocolate bar; and conversely, a $d$-dimensional chocolate bar leads to Divisors played with the product of powers of $d$ primes.

The strategy-stealing argument works fine for Divisors; the first player must have a winning strategy because if the second player did, his winning response to an opening play of "$n$" could itself have been used by the first player as an opening play. But no one knows what that strategy is.

Adventurous folks have considered allowing transfinite ordinals in Chomp. Even more general are "poset games," which begin with a fixed partially ordered set $P$ from which two players alternately choose elements. Neither player may take an element that is greater than or equal to any previously chosen element, and the object is to get the last one. As of this writing, the last nice theorem on poset games was proved by Steven J. Byrnes, a high school senior from West Roxbury, Massachusetts. Steven's theorem won him a $100,000 scholarship in the 2002 Siemens Westinghouse Competition.

## All Roads Lead to Rome

This puzzle had a rather serious origin in R. L. Adler, L. W. Goodwyn, and B. Weiss, "Equivalence of Topological Markov Shifts," *Israel J. Math,* Vol. 27 (1977), pp. 49–63.

After the roads are colored, you can think of $R$ and $B$ as operations on sets of nodes as follows: $R(S)$ is the set of nodes reachable from some node of $S$ by following the red exit, and $B(S)$ similarly. Then the conjecture says that for some coloring, there is a composition of $R$s and $B$s that collapses the full set of nodes to a single node.

The illustration shows two colorings of the complete digraph on three nodes. The first cannot be collapsed, since $|B(S)| = |R(S)| = |S|$ for any $S$. The second, however, is collapsed by $BR$ or by $RB$.

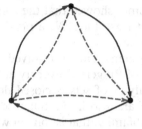 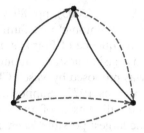

There are some classes of graphs for which the conjecture is known to hold, e.g., if all cities have two roads coming *in*, and the number of cities is odd (see J. Friedman, "On the Road Coloring Problem," *Proc. A.M.S.*, Vol. 110, No. 4 (December 1990), pp. 1133–1135).

### Discs in a Disc

This lovely conjecture is due to Alexander Soifer of the University of Colorado, Colorado Springs. It and its relatives have been the subject of a dozen articles in the journal *Geombinatorics*; it is known, for example, that squares of total area 1 can be packed into a square of total area 2. The generalization to higher dimension was suggested by your author, among others; the case of two balls, each of volume $\frac{1}{2}$, shows that $2^{d-1}$ is best possible.

### The Union-Closed Set Conjecture

We finish with a puzzle about those simplest of mathematical objects, finite sets. Alas, even these lead to fiendish open problems.

This particularly notorious conjecture seems to have arisen first in the 1970s in the work of Peter Frankl, a Hungarian mathematician who lives in Japan (and is a famous TV personality there). It has been driving combinatorialists crazy since then, but as yet they have not even established that there is an element in *any* fraction $c > 0$ of the sets.

A very clever proof of E. Knill, cited in a paper by Piotr Wojcik ("Union-Closed Families of Sets," *Discrete Math.*, Vol. 199 (1999), pp. 173–182), shows that there is an element contained in at least $N/\log_2 N$ sets, where $N$ is the size of the family.

The latest progress on the puzzle was recorded by David Reimer of The College of New Jersey, in *Combinatorics, Probability & Com-*

*puting*, Vol. 12 (2003), pp. 89–93. Reimer showed that the average set size in a union-closed family is at least $\frac{1}{2}\log_2 N$, a previously open consequence of Frankl's conjecture.

Many basic questions about set systems remain unsolved. Another was proposed by Vašek Chvátal of Rutgers University at least as far back as 1972. Suppose a family $\mathcal{F}$ of sets is closed downward, i.e., all subsets of any set in $\mathcal{F}$ are also in $\mathcal{F}$. Suppose you want the largest possible *intersecting* subfamily, that is, one in which any two sets have nonempty intersection. One way of obtaining an intersecting family is to take all sets in $\mathcal{F}$ containing a fixed well-chosen element. Chvátal's conjecture says that you can never do any better.

# Afterword

What you have read is, or was intended to be, a collection of mathematical puzzles—not a book about mathematics. It is devoted to amusing problems, not important ones. It builds no theory, fits in no structure, and imposes no discipline; its attention span is the size of a toddler's.

Even a proponent of the problem-oriented approach to mathematical research (like Tim Gowers, author of an article called "The Two Cultures of Mathematics") would blanch at the notion of learning mathematics from a puzzle book. Your author does not disagree.

Nonetheless, I have a feeling that understanding and appreciating puzzles, even those with one-of-a-kind solutions, is good for you. I have not attempted here to capture a problem-solver's thinking, as Polya and others have done, preferring to let the puzzles speak for themselves. But the puzzles do speak, and they speak the truth.

*—Peter Winkler*
*July 9, 2003*

# Index of Puzzles

Peter Winkler is Director of Fundamental Mathematics Research at Bell Labs, Lucent Technologies. Currently he is on sabbatical at the Institute for Advanced Study, in Princeton, NJ. He is the author of 125 research papers in mathematics and holds a dozen patents in computing, cryptography, holography, optical networking and marine navigation.

In some circles he is best known as the inventor of cryptologic methods for the game of bridge, currently illegal for tournament play in most of the western world.

Printed in the United States
by Baker & Taylor Publisher Services

Printed in the United States
by Baker & Taylor Publisher Services